D0892044

WITHDRAWN

W. J. O'Connor

Founders of British physiology

A biographical dictionary, 1820-1885

A collection of biographical sketches of over a hundred of the most important British physiologists who worked during the heroic period of Victorian medical research.

For the first time, easily accessible accounts of men legendary to modern physiologists have been gathered together in a single volume. The biographies are grouped into the period in which the men worked and into the institutions to which they were attached. The sections and chapters are designed so that the short explanatory texts can be taken together to form a concise history of the development of physiology in Britain during the nineteenth century.

W. J. O'Connor is Honorary Lecturer in Physiology at the University of Leeds.

Manchester University Press

Manchester and New York
*Distributed exclusively in the USA and Canada
by* St. Martin's Press Inc.,
175 Fifth Avenue, New York 10010, USA

Copyright © W. J. O'Connor 1988

Published by Manchester University Press
Oxford Road, Manchester M13 9PL, UK

Distributed exclusively in the USA and Canada
by St. Martin's Press, Inc., 175 Fifth Avenue, New York, NY 10010, USA

British Library cataloguing in publication data
O'Connor, W. J.
  Founders of British physiology: a
  biographical dictionary 1820–1885.
  1. Physiologists – Great Britain –
  Biography 2. Human physiology – Great
  Britain – History – 19th century
  I. Title
  612'.0092'2   QP25

Library of Congress cataloging-in-publication data applied for

ISBN 0-7190-2537-0 *hardback*

Typeset in Great Britain
by Williams Graphics, Abergele, North Wales

Printed and bound in Great Britain by
Biddles Ltd, Guildford and King's Lynn

Contents

Contents

Preface

The nature of this book is stated by the subtitle, *Biographical dictionary*; about three-quarters of the text consists of some 100 biographies of men who can be regarded as physiologists in the years before 1885. My belief that these may form a useful reference book arises from the circumstances which led to its production.

When in 1976 I retired from active work in the Physiology Department in Leeds, I found that I now had time to explore my general interest in Victorian novelists, particularly George Eliot. A physiologist interested in the life of George Eliot must also become interested in George Henry Lewes and this was further stimulated in 1976 by the meeting of the Physiological Society to commemorate the centenary of its foundation. Interest in Lewes involves interest in other physiologists of the period and hence began a search for information about 'founders of Physiology'.

This search could not be easily followed, however, because there is no book describing the development of physiology through the last century. Information about the founders of British physiology is only available about each individual separately, in the form of Obituary Notices in *Dictionary of National Biography*, *Lancet*, *British Medical Journal*, *Nature*, *Proceedings of the Royal Society*. For individuals not included in *DNB*, it was often not immediately easy to find obituary notices, because the date of a man's death is not usually given when his name is mentioned. Thus I began to see that a biographical dictionary would have the value of providing easily accessible accounts of men whose names are known to modern physiologists only by the somewhat scanty historical notes in many textbooks of physiology. By means of personal contacts and informal seminars in the Department of Physiology in Leeds, I found most of my colleagues interested in the men of the last century but doing nothing about it. With compact biographies available, physiologists should be helped towards knowledge of our scientific origins.

The biographies in this book are drawn from published obituary notices and I have not pursued matters further by research into the archives of institutions or societies: this is work for historians rather than an interested physiologist. Historians distrust obituary notices; written by a friend of the recently deceased, they contain no criticism and, by the custom of the nineteenth century, are often absurdly adulatory. Where historians have researched into the archives, a more realistic picture often emerges. On the other hand, the obituary notices do reveal the contemporary opinion of the man's contribution to physiology and so indicate how his immediate successors were influenced by his work.

Obituaries give reliable facts about the sequence of appointments in a man's career and his contacts with other men of his time – his teachers, colleagues and pupils. These are basal facts; I have tried to avoid drawing historical deductions from them. In the manner of physiological argument, I would say that these are the facts on which theories about the influence of factors could be based, but I do not attempt to present such theories. In grouping the biographies into Parts I, II, III and into chapters with brief introductions, an account of the development of physiology inevitably emerges, but I have not attempted to argue the story.

The year 1885 is a somewhat arbitrary choice of the time when the process of founding physiology ended and this was partly determined by the fact, important to a physiologist in Leeds, that a Professor of Physiology was appointed here in 1884. This was one of many appointments by which about that time the major British Schools of Physiology were firmly established; after 1885 the focus of interest in our physiological ancestors changes to the work they did, rather than the institutions which they founded. The Physiological Society, founded in 1976, had 48 members in 1880. The subjects in Part I of the book lived before the foundation of the Physiological Society; most of those in Part II survived until the Society was formed and became connected with it; while those in Part III all joined the Society at the beginning of their careers. This book includes notices of all members who joined the Physiological Society before 1880. The notices extend beyond 1885, however, in that they give accounts of the entire life of those physiologists who joined the Society about 1880 and sometimes continued to work in the subject until 1930. Sharpey-Schafer was one such example and I must pay tribute to the mass of information about early members contained in his *History of the Physiological Society during its first fifty Years, 1876–1926* (1927, Cambridge University Press).

In the text, smaller type is used to distinguish the biographies from the brief paragraphs describing the general background of each time and institution. This corresponds with two aspects of the history of physiology in this period. On the one hand there was general development which was related to the social history of the time and would have occured irrespective of the individuals who did in fact actually bring it about. On the other hand, it was the characters of the 'Founders of Physiology' which determined precisely when and how each development did in fact occur. As an example, after 1870 the use of anaesthetics made it certain that animal experimentation would become the method of physiologists and at that time also teaching by practical classes was becoming generally used in all scientific subjects. To what extent was the place and time of these developments in England determined by the characters of Sharpey, Burdon Sanderson and Foster? I have been struck by the coincidence which brought three such extraordinary personalities together. In this book, the general background is briefly indicated; the characters of the men are given fully in the biographies.

In preparing the book for publication, it was decided not to print portraits. In nearly every case, a portrait is provided with the obituary notice and one could only reproduce this same portrait. In the lists of references the letter P indicates that the obituary notice contains a portrait.

I gratefully acknowledge encouragement from the staff of the Department of Physiology, University of Leeds, and I particularly thank Mrs J. Hill, who typed the manuscript.

To my wife

PART I:
PHYSIOLOGISTS 1820-35

1

Anatomical physiologists

In September 1826, the *Lancet* (Vol. 9, pp. 18–30) began what was to be an annual item of great value to historians by publishing for the first time a list of lectures available to medical students in London. For the session 1826–7, a course in 'anatomy, physiology and surgery' was available at four hospitals (St Bartholomew's, St Thomas's, Guy's, The London) and at seven private schools (Webb Street, Blenheim Street, Dean Street, Great Windmill Street, Little Windmill Street, Chapel Street and Berwick Street). The names of the lecturers included Abernethy at Barts, Cooper and Blundell at Guy's, Grainger at Webb Street, Bell at Great Windmill Street and Mayo at Berwick Street. In the 1820s the private medical schools had many more students than attended the courses at the hospitals (Newman, 1957) and Bell, Grainger and Mayo were regarded as the best teachers in anatomy. These lectures were given daily and the fee for the whole course was about five guineas. Only one school offered lectures specifically in physiology. At Guy's Hospital, Dr Blundell gave a course of two lectures per week (fee £2 2s) on 'physiology or laws of animal economy'.

The schools also provided a dissecting room for the students, arranged demonstrations on dissected preparations and had extensive museums of anatomy and pathology. Primarily these were courses in the anatomy of the dissecting room, applied to surgery. The limited 'physiology' contained in them can be assessed from the following number of the *Lancet* where the opening lecture in each school was published. Amongst these men some, such as Bell, Mayo, Grainger and Blundell, did some experimental work of a physiological nature; the name of Bell is known to modern physiologists in connection with the function of the spinal nerve roots.

Mazumdar (1983) refers to these teachers as 'anatomical physiologists' and examines their dominant influence in medical teaching in the 1820s and their decline after 1830.

Similarly in the provincial medical schools and in Edinburgh, medical education in 1830 was dominated by surgeons. When the medical school in Leeds opened in 1830, lectures in 'anatomy, physiology and pathology' were given by T. P. Teale (Senior), J. P. Garlick and C. T. Thackrah. Teale and Garlick were young surgeons with no particular interest in physiology. Thackrah (1795–1835) had previously run a private school of anatomy, given popular lectures in physiology and published works on the blood and on social conditions. He was one of the first in England to concern himself with the effects of working conditions in factories and can be regarded as a pioneer in many matters of public health.

In Edinburgh in 1830 the Professor of Anatomy was Alexander Monro, the third of the family who had taught anatomy and surgery and the least effective of them as a teacher. Far more students, about 400, attended the lectures of Barclay and Knox in their extra-academic school. In Edinburgh some Physiology was taught apart from Anatomy, in the lectures of the Professor of the Institutes of Medicine (see p. 14).

Sir Charles Bell, FRS (1774–1842)

Charles Bell was born in Edinburgh. His father, a minister of the Episcopal Church, died when he was only five and Charles's early life was directed by his mother. There were three older brothers who, with the support of Sir Walter Scott, became eminent in Edinburgh – John as a surgeon, George as Professor of Law, and Robert, writer to the Signet. Charles made little of his formal early education but became associated with the painter David Allan, thereby developing the natural aptitude for drawing which was an important factor in his subsequent career.

In 1798–99 Charles was a medical student in Edinburgh, qualifying as a Member of the Royal College of Surgeons, Edinburgh. During these student years he produced a book, *A system of dissections* which became widely admired for its beautiful illustrations. Elected FRCSE, Charles became surgical attendant at the Edinburgh Infirmary and with his brother John, established as a surgeon, produced a book on *The anatomy of the human body* which included a section on the nervous system.

A career in Edinburgh became impossible, however, because John was in serious disagreement with senior medical men in that city, and so in 1804 Charles went to London, walking the last sixty miles from Huntingdon because there were no mails forward on Saturday and Sunday. He arrived with very little money and no assured position, but with introductions from well-known people in Edinburgh.

He was immediately accepted by surgeons because of his *System of dissections*, and by artists because of his book *The anatomy of expression* which appeared in 1806. This acceptance did not produce much income, so he took a decrepit house in Leicester Street, Leicester Square, in which he provided lodgings and lectured on anatomy to artists and to medical students. He also entered private surgical practice, and these enterprises prospered. Charles Bell's lectures on anatomy and surgery were advertised to be given at 10–11 am every day at Leicester Street; the dissecting room was open from 8 am – 10 pm, with demonstrators always in attendance. In 1810, he extended his activity as a private teacher by purchasing a share in the Great Windmill Street School of Medicine, founded forty years earlier by Hunter. By adding his collection of specimens and models to that already in the School, Bell formed a 'noble museum' which was later sold to the Edinburgh College of Surgeons.

In addition to his activity as a teacher, Bell developed skill and a practice as a surgeon. After the Battle of Corunna (1809) he went to the south coast to study and draw the wounds of the casualties as they were brought back, but found there was surgical work still to be done amongst them and devoted himself to it. Similarly when news arrived of the battle of Waterloo (June 18–19, 1815) he immediately set off to record the bone injuries, but reaching Brussels on 30 June, he then spent three days without sleep giving surgical treatment, particularly to French casualties. Letters written from Brussels to his brother George in Edinburgh were shown to Sir Walter Scott, who 'was set on fire' by them and on 27 July set off for Brussels and Paris. From this journey Scott wrote *Paul's letters to his kinfolk*; later he wrote his book *History of Napoleon*. Throughout his life Charles wrote letters to his brother George and the *Letters of Sir Charles Bell*, published in 1870, are an important human document of the period, also telling us much about Charles Bell himself.

Bell gained high standing in private practice and as a teacher in his private school at Great Windmill Street but, until 1813, had no appointment to a London Hospital. He then became surgeon to the small Middlesex Hospital which in the next few years he developed into a medical school. In 1824 he was appointed Professor of Anatomy and Surgery at the Royal College of Surgeons, where his lectures attracted large and enthusiastic audiences. By now well-known, he was active with Lord Brougham in the foundation of the University of London in which he was one of the original professors, and on 1 October 1828 he gave the first lecture in the University, describing himself as Professor of Physiology and Surgery. For the next two years he gave two lectures per week in Physiology until disagreement amongst the professors caused his resignation: he continued to lecture in the Middlesex School, without fees. In 1824 he sold the Great Windmill Street School to Mayo. However in 1830 the School closed because of its close proximity to the Middlesex School and the University of London (University College).

In 1836 Bell was elected Professor of Surgery in the University of Edinburgh and exchanged his immense degree of activity in London for a quieter life, still lecturing and in surgical practice, but now with time for country excursions and the delights of angling. He suffered from angina pectoris and died in such an attack while on holiday at Hallow Park, near Worcester. He is buried, with a tablet to his memory, at Hallow Church.

Knighted in 1831, Sir Charles Bell was famous as a brilliant surgeon and teacher of Anatomy and Surgery. Amongst other writings three books represent his professional life: *System of dissections* (1798); *A system of operative surgery founded on the basis of anatomy* (1807); *Essays on the anatomy of expression in painting* (1806). This last book was written for artists but is of interest to anatomists, and aptly shows the contribution of his skill in drawing to his professional career. Finally he co-operated in producing a new edition (1826) of his brother John's book *The principles of surgery*, making this available in a cheaper form for more general use.

During his active career as surgeon and teacher, Bell privately thought and worked as a physiologist. In 1819 he published an *Essay on the forces which circulate the blood*, in which he maintained that one of these forces is contractions of the arteries. But his main interest was the nervous system. He first provided anatomical description (*Anatomy of the brain explained in a series of engravings*, 1802; *A series of engravings explaining the course of the nerves*, 1803). From about 1807 he was thinking about function and in 1811 wrote out his ideas in *Idea of a new anatomy of the brain: submitted for the observation of his friends*. About 100 copies were printed and distributed; it was reprinted in 1868 in the *Journal of Anatomy and Physiology*, Vol. 3, with notes on the circumstances of this work. From 1821 Bell communicated papers to the Royal Society and these were collected, with some modifications, in *The nervous system of the human body* (1830, 1836, 1844). Bell himself regarded this work on the nervous system as the major original contribution of his life and it has been accepted by physiologists as the first suggestion that different parts of the nervous system subserve individual functions.

For example, Bell noticed that touching the optic nerve (or lateral pressure on the eyeball) caused the subject to see a flash of light and so he defined parts of the brain as subserving vision. Bell argued that the ability to weigh an object in our hands requires that there is sensory innervation of muscles. By this kind of argument, Bell was led to a general thesis that different parts of the nervous system subserve different functions; for example he speculated on the sensory or motor functions of the cerebellum and cerebral cortex. As a small part of his *Idea of a new anatomy* he indicated that the anterior nerve roots were motor and the posterior sensory, each with its own related part of the spinal cord. Bell did some experiments but not many, and it was Magendie who made experimental tests of the functions of the nerve roots.

Bell argued from anatomy rather than experiments, following the main tracts by gross dissections; he did not use a microscope. In 1821 his ideas were conveyed to Magendie in Paris by his assistant, friend and brother-in-law, John Shaw. Soon after, in 1822, Magendie demonstrated by direct experiments on dogs the separate function of the two roots. Unfortunately the priority of the two men became the subject of a famous controversy, with Bell's claim asserted by Bell himself and John Shaw. When the arguments are clearly set out, as by Olmsted (1944) in his study of François Magendie, it is surprising that Bell involved himself in the rather futile debate about priority. Perhaps this was to justify the standing of the old-style anatomical physiologist, Bell, against the new-style experimental physiologist, Magendie. In passages from his letters, cited by Olmsted (1944), Bell expressed the outlook of the anatomical physiologist:

Experiments have never been the means of discovery; and a survey of what has been attempted of late years in physiology will prove that the opening of living animals has done more to perpetuate error, than to confirm the just views taken from the study of anatomy and natural motions... The whole history of medical literature proves, that no solid or permanent advantage is to be gained, either to medical or general science, by physiological experiments unconnected with anatomy.

Perhaps also there was a personal dislike for vivisection experiments. In a letter to his brother George in 1822, Bell said:

I should be writing a third paper on the nerves, but I cannot proceed without making some experiments which are so unpleasant to make that I defer them. You may think me silly but I cannot perfectly convince myself that I am authorised in nature or religion to do these cruelties – for what? – for anything else than a little egotism or self aggrandisement.

Such doubts persisted in the minds of English physiologists for many years, indeed until about 1875 when effective anaesthetics made animal experiments entirely acceptable (see Part III).

Herbert Mayo (1796–1852)
It appears that about 1820 Mayo became asistant to Bell at the school in Great Windmill Street. In the *Lancet* of 1826, Mayo is listed as giving a course in anatomy, physiology and surgery at a private school in Berwick Street, but in the same year Mayo bought the Great Windmill Street school from Bell (p. 5). However, Great Windmill Street was inevitably losing its students to the nearby Middlesex and University College and it closed in 1830; Mayo was then appointed Professor of Anatomy and Physiology at King's College at the time of its foundation. Mayo seems not to have been a success there; staff relations were mixed and it seems that Mayo was not a good lecturer (Hearnshaw, 1929). In 1836 Mayo left King's to become a full-time surgeon at the Middlesex Hospital, where he remained until 1844, when he went to live in Germany.

An obituary notice in the *Medical Times and Gazette* states that increasing rheumatism first sent him to Germany to seek cure by hydropathy. Although a cripple with major joints ankylosed, Mayo became a partner and medical adviser to a hydropathic establishment in Germany.

In a less kind obituary notice, the *Lancet* briefly states 'he forsook the legitimate path of his profession and became a mesmerist and hydropath'. He practised in Germany as a hydropath and died there.

About 1822 Mayo published physiological papers on the function of the nervous system including the function of the anterior and posterior spinal nerve roots. His relationship with Bell in this work is uncertain. Mayo had earlier been assistant to Bell and he was given a copy of *Idea of a new anatomy of the brain* (p. 5). Later the two men became antagonistic. The controversy between Bell and Magendie obscured the contributions made by Mayo and others about 1820 and Mayo involved himself by arguing for his own contribution. After he had left King's College and was working as a surgeon at the Middlesex, Mayo published in 1839 a book *Outlines of human physiology*. This book evoked a biting review in the *Lancet* in which he is accused of altering dates of references he cites, to make it appear that some of his work on the spinal roots preceded that of Bell and Magendie. 'Upon this thimble-rigging manoeuvre rest the scientific claims of Mr Herbert Mayo. We will not stoop to criticise it' (*Lancet* 1839–40, 38, 342–6).

Richard Dugard Grainger, FRS (1801–1865)

Richard and his older brother Edward were both born in Birmingham. Edward studied surgery at Guy's under Astley Cooper and then began a private school in nearby Webb Street. Edward turned out to be a good teacher and by 1820 his school in Webb Street had grown to have some 300 students. Richard began a career in the Army by entering as a cadet at Woolwich but was persuaded to join his brother in the Webb Street school, teaching anatomy, physiology and surgery. Edward was already ill with tuberculosis and died in 1824 leaving the young Richard in full charge of the school. Richard was apparently not as good a teacher as his brother but was sufficiently able and liked by the students. The Webb Street school maintained its position until it declined along with the other private schools as medical eduction passed to the hospitals. The Webb Street school closed in 1842 when Richard Grainger was invited to become lecturer in Anatomy and Physiology at St Thomas's Hospital Medical School. He continued to lecture at St Thomas's until he retired in 1860 because of poor health. On his retirement he refused any personal gift but his friends and ex-pupils founded a Grainger Testimonial Prize for an essay on a physiological topic.

While at St Thomas's Grainger was also inspector to various matters of public health, such as burial grounds, and he also performed much charitable work. He published a report on the cholera epidemic of 1848–9 and other reports on sanitation and workers' welfare. Grainger was a devout member of the established Church and in his private and public life a much loved man.

Early in his career Grainger published a book *The elements of general anatomy, containing an outline of the organisation of the human body* (1829), which was much used by students of that time. General anatomy was the term used to cover what was then taught about the make-up of organs, following the ideas of Bichat (see p. 12); this was the physiological part of the lecture courses about 1830. Also while still at Webb Street, he made original studies of the spinal cord and in 1837 published *Observations on the structure and functions of the spinal cord*, which gave anatomical basis for some of Marshall Hall's ideas on the function of the spinal cord (p. 17). He was elected FRS in 1837.

James Blundell (1790–1878)

James Blundell, born in London, was privately educated in the classics before entering medical education at the Southwark United Hospitals (St Thomas's and Guy's). He lived with his uncle, Dr Haighton, who was Lecturer in Midwifery and Physiology at Guy's. He then went to Edinburgh, where he graduated MD in 1813.

He returned to London in 1814, becoming joint Lecturer in Midwifery with his uncle, Haighton, and about 1818 commenced a private course of lectures in physiology and midwifery near the hospitals. After being admitted Licentiate of the Royal College of Physicians in 1818, he became Lecturer in Physiology and Midwifery at St Thomas's and Guy's in succession to this uncle. Over the next fifteen years he became a great figure as a teacher and practitioner in obstetrics and also as a physiologist. Blundell's lectures were unique in that they were specifically on physiology apart from anatomy. Their full title was 'Physiology, or Laws of Animal Economy' and the course was two lectures per week at a fee of £2 2s. These lectures were handed on to him by his uncle and had their origin in circumstances in the career of Haighton (see below).

In 1834 Blundell was in a violent disagreement with the treasurer of Guy's Hospital, claiming that a co-lecturer in obstetrics had been appointed without his knowledge or approval. Blundell regarded the lectureship as his personal property, only held in the hospital as a matter of convenience and so outside any jurisdiction of the hospital authorities. In protest Blundell resigned amid a somewhat terse correspondence in the *Lancet* (Vol. 27, 1833–4). Thereafter he did not lecture but had a large private practice despite his eccentric custom of not rising until about noon, seeing his consulting room patients until 6 pm, dining and then going in his yellow carriage with a lamp to let him read, to visit patients in their homes, often as late as midnight. He was very highly regarded by his contemporaries and in 1838 was the subject of a biography in Pettigrew's *Medical Portrait Gallery*, published while he was still in active practice. His writings on obstetrics included a large course of lectures at Guy's published in the *Lancet* in 1827–8, and later published as a book *The principles and practice of obstetrics*.

In Physiology Blundell was a true experimenter, particularly in three topics. Firstly, by experiments on rabbits he demonstrated that opening and handling the peritoneum was not so damaging as was then believed and thus he advocated and performed intra-abdominal surgery, such as removal of a cancerous uterus, with some survivals. Secondly he observed that the *corpus luteum* could exist under circumstances where conception could not have taken place. Thirdly, by experiments on dogs he showed that animals could survive the transfusion of blood of their own but not other species. He recommended, and in some cases used, blood transfusion for the treatment of serious postpartum haemorrhage and other conditions; about half of his own cases survived. This work, carried out in 1818–23, formed three parts of a book published in 1825, *Researches, physiological and pathological: instituted with a view to the improvement of medical and surgical practice.* He was a firm advocate of experiments on animals against the criticisms and false representations of antivivisectionists.

Retired from practice, he lived to the age of 88, but apparently did no further physiological work after his early years. He accumulated many books which he left to the Obstetrical Society of London.

John Haighton (1755–1823)
Blundell's uncle, John Haighton had been a pupil at St Thomas's and became demonstrator in anatomy under the surgeon, Herbert Cline. However Cline also had a favourite pupil in Astley Cooper, and offered a lectureship to this younger man, over Haighton's head. In 1789 therefore Haighton started his own course of lectures at the United Hospitals on physiology and midwifery and continued to give these lectures until 1818 when he handed them on to his nephew Blundell. Manuscript notes of his lectures of 1796 are preserved at Guy's Hospital. Blundell described his uncle as 'a man of kindest heart, generous and with a remarkable regard for the sanctity of truth. He was a little irritable but it was only a hasty spark ... folly used to vex him; he would not laugh at her ups and downs'. The same might be said of Blundell himself.

Haighton even at that time was regarded as a rather ruthless experimenter on animals and published papers on vomiting (1789), 'reproduction' (restoration of function) in divided nerves (1795) and impregnation in many species of animals (1797). In succeeding his uncle as lecturer in physiology, Blundell was following a man who had established the idea of experimental investigation.

The relations of Haighton and Blundell to the hospital are described by Cameron (1954).

Research by anatomist-surgeons

Bell, Mayo, Grainger and Blundell are examples of the anatomist-surgeons who dominated medical teaching up to 1830. There were many others, some of great fame as surgeons, who did not contribute to physiology. Both Astley Cooper (1768-1841, St Thomas's and Guy's) and John Abernethy (1764–1831, St Bartholomew's) retired in 1827 after giving lectures in anatomy and surgery for forty years.

Anatomist-surgeons such as Bell, Mayo, Grainger and Cooper often carried out research which might be of a physiological nature. Because the medical schools had no laboratories this was done mostly in their own homes. Astley Cooper maintained a large private dissecting room in an old warehouse at his home in St Mary Axe (Brock, 1952) where he taught students and dissected very actively on human subjects and various animals, including fish obtained from the nearby Billingsgate market. He also had an arrangement by which he dissected any animals which died at the menagerie of the Tower. This provided an elephant which, being too large to go into the dissecting shed, was completely dissected in the yard behind tarpaulins. A close friend of Cooper's was Edward Coleman (1765–1839), who in 1793 became Professor of Anatomy at the Veterinary College and was largely responsible for the development of that institution. Later, under the influence of Cooper, the Duke of Kent became patron and the college became the Royal Veterinary College. Coleman in his student days had become interested in asphyxia and in 1891 published *A dissertation on suspended respiration from drowning, hanging and suffocation: in which is recommended a different mode of treatment from any hitherto pointed out:* this gained a medal of the Humane Society.

Work published by Cooper was anatomical or surgical. It included anatomical description of the thymus and experiments in which the femoral artery of the dog was tied to investigate the anastomatic channels. With physiology as yet an undefined subject, it was rather accidental that the private research work of Bell, Mayo and Grainger can now be regarded as contributing towards our understanding of the nervous system, particularly the function of the spinal cord.

Teaching by anatomist-surgeons

In the period before 1830 the basis of medical education was the dissecting room. The private schools all originated as schools of anatomy, with the owner and director of the school giving the lectures in anatomy, organising the dissecting room and accumulating a museum. The teaching of anatomy was directed towards surgery and was often better in the private schools than in the hospitals (Newman, 1957). This was due partly to the higher prices that the private schools would pay to 'resurrectionists' who provided the bodies.

There seems to have been a state of neutrality by which the anatomists were not prosecuted, until the scandal of the Burke and Hare case in Edinburgh led to the Anatomy Act of 1832. Thereafter the hospital schools received enough bodies and one of the advantages of the private schools was lost.

When the University of London began in 1828, medical subjects were included because medical students were by far the largest group who would be attracted to the lectures and so provide income for the new University. The inaugural lecture in the University was given by Bell (p. 4). By 1830 there were two colleges of the University: University College, London, and King's College; following their pattern and in accordance with the developing educational ideas of the time, the Hospitals began to be more concerned with teaching and to give it better organisation.

In 1826 Bell was the most experienced teacher of anatomy in London and was appointed as one of the first group of Professors in the University of London. However he was not the only professor and was expected to share the teaching of anatomy with others and was saddled with a senate who were thinking of education in terms other than the direct teaching of professional skill, which was the expertise of the anatomist-surgeons such as Bell. Mazumdar (1983) has recently studied the correspondence between Bell and the administrators before Bell resigned in 1830. Similarly Mayo was appointed Professor in King's College in 1830, was not successful, and resigned in 1836. In Guy's Hospital after 1807, the treasurer, Harrison, was bringing organisation into the teaching and it was in a row with Harrison that Blundell gave up his lecturership in Physiology and Midwifery (p. 9).

About 1830 can be taken as the time when medical education passed out of the dominance of the surgeon-anatomists. For many years yet, the lectures in physiology remained linked to anatomy but surgery was no longer such a dominant interest.

The little physiology that was taught was often called 'general anatomy' to distinguish it from the main topographical anatomy. General anatomy was largely speculative and based on the ideas of Bichat (1771–1802) whose brief career in Paris is regarded as having initiated the proper study of histology and pathology. Bichat believed that the different parts of the body were composed of the same few tissue elements arranged in different proportions and in different relation to each other. In 1820 this was still hypothetical because microscopes were not yet capable of direct description of tissues. 'General anatomy' contained no microscopic descriptions and no experimental work. Its arguments had then little relevance to surgery and have now little resemblance to modern physiology. One reason why the teaching of physiology passed out of the province of the anatomist-surgeons was that the special techniques of the microscope required special knowledge (see Chapter 4).

References

Bettany, G. T. (1885). *Eminent Doctors: their lives and their work.* 2 vols. London: Hogg.

Brock, R. C. (1952). *The Life and Work of Astley Cooper* (P). London: Livingstone.

Cameron, H. C. (1954). *Mr. Guy's Hospital 1726–1948.* London: Longmans Green.

Hale-White, W. (1935). *Great Doctors of the Nineteenth Century.* London: Edward Arnold.

Hearnshaw, F. J. C. (1929). *The Centenary History of King's College, London.* London: Harrap.

Mazumdar, P. M. H. (1983). 'Anatomical physiology and the reform of medical education: London, 1825–1835', *Bulletin of the History of Medicine,* **57**, 230–46.

Newman, C. (1957). *The Evolution of Medical Education in the Nineteenth Century.* London: Oxford University Press.

Olmsted, J. M. D. (1944). *François Magendie.* New York: Schuman's.

Pettigrew, T. J. (1838–40). *Medical Portrait Gallery: Biographical Memoirs of the most celebrated Physicians, Surgeons, etc.* 4 vols. London: Fisher.

Biographies

Abernethy: Pettigrew, Vol. 2 (P).
           Bettany, Vol. 1, 226–41.

Bell: Pettigrew, Vol. 3.
           Bettany, Vol. 1, 242–63.
           Dictionary of National Biography, **4**, 154–7.
           Hale-White, 42–62.

Bichat: Pettigrew: Vol. 1 (P).

Blundell: Pettigrew, Vol. 1 (P).
           British Medical Journal (1878), i, 351–2.

Cooper: Bettany, Vol. 1, 202–26.
           Pettigrew, Vol. 1 (P).
           Hale-White, 22–41.
           Brock (1952).

Grainger: *British Medical Journal* (1865), i, 176.
           Lancet (1865), i, 190–1.

Haighton: Cameron (1954), p. 160.
           Dictionary of National Biography, **23**, 441–2.

Mayo: *Lancet* (1852), ii, 207.
           Medical Times and Gazette (1852), **5**, 226.

2

Physicians 1820–35

Physicians were of relatively little importance in hospitals in about 1830, the dominating influence being the surgeons, as described in the previous chapter. However at this time physicians began accurately to describe disease conditions and in so doing provided some beginnings of modern physiology. Two physicians of Guy's Hospital, Bright and Addison, contributed to physiology in this way: Addison is particularly known to modern physiologists by 'Addison's disease' describing the effects of loss of function of the adrenal glands.

In Scotland this scientific role of the physician was given formal recognition by the appointment of a Professor of the 'Institutes of Medicine', sometimes called the 'Theory of Medicine' or 'Physiology'. The Professor of the Institutes was required to work in the hospital and in fact the chair was something of a junior chair in medicine, junior to the Professor of Clinical Medicine. The professor of the Institutes of Medicine in Edinburgh in 1822–42 was Alison (see below) and he contributed little to physiology before he passed on to become Professor of the Practice of Medicine. A successor, Hughes Bennett (p. 96), introduced teaching in the physiology of his time and finally Rutherford (p. 198) undertook no clinical work and adopted the title of Professor of Physiology.

By far the greatest physician-physiologist of the 1830s was Marshall Hall who was an active consultant in London without hospital or teaching appointments. He chose to devote himself to experiments in physiology, often with direct application to medicine and Marshall Hall and Charles Bell can be regarded as the first British physiologists.

William Pulteny Alison (1790–1859)

Alison was born at Boroughmuirhead, near Edinburgh, the son of a clergyman who was later in charge of the episcopalian congregation of Edinburgh. His mother was the daughter of Dr John Gregory and so he was connected with a family having long association with the

University of Edinburgh. William Alison was educated privately and entered Edinburgh College in 1803 to study arts and medicine, as his father wished, although he himself might have preferred the army. He passed MD in 1811. During his academic career, he studied under Dugald Stewart and so had a deep interest in philosophy and indeed might have been Stewart's successor in the chair of philosophy.

He began medical work by becoming physician to the newly-founded New Town Dispensary, where he gained deep sympathy with working-class people. In 1820 Alison was appointed Professor of Medical Jurisprudence in Edinburgh and in 1822 changed to be Professor of the Institutes of Medicine, holding this post until 1842 when he was promoted to the senior chair of Professor of the Practice of Medicine. In these chairs he served as a physician at the Infirmary and acquired at the same time a large private consulting practice. He was forced by attacks of epilepsy to resign his chair and practice in 1856 and he died in 1859 at his home near Colinton. He became vice-president of the British Medical Association and presided over its annual meeting held in Edinburgh in 1858.

The substance of his lectures in the Institutes of Medicine was collected into a book *Outlines of physiology* published in 1831. The book promoted the idea that there is a 'life force' superadded to the physical forces of dead matter but it contained no account of experimental work. The later part of Alison's life was devoted to ameliorating the lot of the poor. From his experience with poor patients Alison formed the idea that poverty, particularly in epidemics of cholera and other fevers, had been a big factor in the spread of diseases and that improvement in living conditions was necessary. He was deeply involved in arguments about how this was to be achieved.

Marshall Hall, FRS (1790–1857)

Marshall Hall was born at Basford, Nottinghamshire. His father, Robert Hall, was a cotton manufacturer who first used chlorine to bleach cloth and his brother, Samuel, also contributed to the bleaching process and suggested the use of a condenser to recover the water used in steam engines. Marshall Hall was for a time assistant to a practitioner in Newark, before he proceeded to Edinburgh University in 1809, graduating Doctor of Medicine in 1812 and becoming House Physician at the Royal Infirmary. In 1814, on foot and armed, he travelled on the continent, visiting Paris, Berlin and Gottingen, and then in 1815 he began to practise as a physician in Nottingham.

In 1826 he moved to private practice in London until his retirement in 1852. He then visited America but suffered from dysphagia which progressed until he could not swallow any food and he died in 1857. By his own request, his case and the post mortem findings were fully described in both the *Lancet* (1857, ii, 253) and the *British Medical Journal* (1857, 760); there was a complete stricture of the lower end of the oesophagus.

Hall held no appointments to hospitals but became recognised as a great consultant physician. In comparison with others of his period, he did not lecture much. He lectured in medicine at the private schools in Aldersgate Street (1834–6), Webb Street (1837–9) and Sydenham College (1837–9), and later at St Thomas's Medical School (1842–6) but one of his biographers said he suffered from 'clergyman's sore throat' which perhaps explains why he lectured so little. We can assess his clinical work by his books. Within five years of qualifying he wrote *On diagnosis* (1817) which introduced to England the scientific clinical approach of Laënnec and Louis in Paris (p. 22). It was this book which brought him his first recognition when he moved to London; it was a remarkable work by one so young. He also wrote other books on clinical topics, including one on the *Diseases of females* (1827) in which he states the necessity of exercise in the open air in female youth. His total clinical experience was collected when he published in 1837 *Principles of the theory and practice of medicine*, which was a book for students. By profession he was a much respected and prosperous consultant physician, who remained apart from the politics of the medical world.

Throughout his career, Hall was privately carrying out research in physiology in his own home, on topics often directly relevant to medical practice. In 1830, he published *Observations on blood-letting, founded upon researches on the morbid and curative effects of loss of blood*; in this book he makes observations on the effects of withdrawing different amounts of blood. This was the first step towards rationalising the use of bleeding as a therapeutic measure but he did not go as far as to condemn it entirely. In 1831, Hall published *A critical and experimental essay on the circulation of the blood*. Using a microscope he clearly described the branching of arteries into arterioles and capillaries and their reunion to form venules and veins, in various tissues of cold-blooded animals. Next he turned his attention to the nervous system, describing very effectively the reflex movements produced by stimulation of sensory nerves and listing the five essentials of a reflex — an excitement, a nerve to the centre, integrity of the centre, a motor nerve to the muscle, the contraction of the muscle. This work was presented in papers to the Royal Society in 1833 and 1837 and was incorporated in two books: *Memoirs of the nervous system* (1837, 1843) and *Lectures on the nervous system and its diseases* (1836). Although some of his papers were rejected by the Royal Society, of which he was elected Fellow in 1832, the value of his work was immediately recognised by Sharpey and others.

His last work was a paper in the *Lancet* (1856, i, 393–4) entitled 'Asphyxia; its rationale and its remedy' in which he described a method of artificial ventilation. The method consisted of immediately placing the subject in the prone position, so that the tongue would fall forwards and clear the airway, and then rolling the subject on to one side and back to the prone position. By these manoeuvres applied to a corpse he measured a tidal volume of thirty cubic inches. He set out his plan for the treatment of an apparently drowned person — the immediate beginning of artificial respiration and keeping the subject cool. At that time the recommendations of the Royal Humane Society were to get the subject in front of a fire, with no mention of artificial ventilation. The last year of his life, when he was an invalid, was given up to this controversy in the *Lancet*, by which eventually the Royal Humane Society was forced to issue his method as their official recommendation.

Although most of his own experimental work was on frogs, Marshall Hall was a target of antivivisectionists and felt the need to justify animal experiments. In those days when human surgery was without anaesthetics, Hall could see the need for painful experiments on animals but he could equally regret this necessity. He suggested that there should be a society of physiologists: its function would be to receive any proposal for animal experiments, discuss it, assess its necessity and ensure its proper execution. This proposal was impracticable and when a Physiological Society was formed forty-five years later, it was quite different from the institution proposed by Marshall Hall.

In his book on the circulation in 1831, Hall put forward principles which should guide research in physiology. They were repeated in the *Lancet* (1847, i, 58–60) with the separate title 'On experiments in physiology, as a question of medical ethics' by Marshall Hall, MD, FRS. They are reproduced almost in Hall's own words:

It has become necessary to re-state these principles, from the fact, that certain writers are still to be found cowardly and base enough to calumniate the physiologist, who is guided by the strictest principles of morality and of conscience in his difficult and painful career.

The Principles of Investigation in Physiology   The sources of our knowledge in physiology, as in all natural science, are observation and experiment. The former consists in a sustained and watchful attention to events which pass under our eye in the ordinary course of nature; the latter, in devices for placing natural objects in new or unusual circumstances or situations.

Unhappily for the physiologist, the subjects of the principal department of his science, that of animal physiology, are sentient beings, and every experiment, every new or unusual situation of such a being, is necessarily attended by pain or suffering of a bodily or mental kind. Investigations in this science should be regulated by peculiar laws.

The *first principle* to be laid down for the prosecution of physiology is this: – We should never have recourse to experiment, in cases in which observation can afford the information required.

As a *second principle* of the prosecution of physiology, it must be assumed that no experiment should be performed without a distinct and definite object, and without the persuasion, after the maturest consideration, that that object will be attained by that experiment, in the form of a real and uncomplicated result.

It must be admitted as a *third principle* in physiological investigations, that we should not needlessly repeat experiments which have already been performed by physiologists of reputation. If a doubt respecting their accuracy, or the accuracy of the deductions drawn from them, arise, it then, indeed, becomes highly important that they should be corrected or confirmed by repetition. This principle implies the due knowledge of what has been done by preceding physiologists.

If it has been concluded that a given experiment is at once essential and adequate to the discovery of a truth, it must next be received as an axiom, or *fourth principle* that it should be instituted with the least possible infliction of suffering.

In all cases the subject of the experiment should be from the lowest order of animals appropriate to our purpose as the least sentient ... The batrachian reptiles are especially animals of this kind; and for many physiological purposes may be promptly deprived, not only of sensation, but of motion, by the prompt division of the spinal marrow.

Lastly it should be received as a *fifth principle* that any physiological experiment should be performed under such circumstances as will secure a due observation and attestation of its results, and so obviate, as much as possible, the necessity for its repetition.

Having ascertained any fact or facts, they should, in my opinion, in accordance with a *sixth principle*, be laid before the public in the simplest, plainest terms. Controversy can be of little service to science.

In quoting the opinion of other authors, I think it should be in their own words. From the frequent misconstruction of the meaning of an author, without this precaution, I would consider this as a *seventh* and *final principle* of treating physiology.

Richard Bright, FRS (1789–1858)
Richard Bright was born in Bristol, the son of a prosperous banker; his older brother became member of Parliament for Bristol. Richard went to school in Bristol and then in Exeter, where a school companion was Henry Holland (1788–1873), later Sir Henry, a famous traveller, writer and fashionable London doctor, and cousin to Mrs Gaskell, the novelist. Richard Bright in 1808 went to Edinburgh University, studying firstly general philosophy and then in 1809 anatomy under Monro and Gordon.

In 1810 he was attracted away from his medical education to join an expedition to Iceland proposed by MacKenzie, on which a second junior member was Holland; this was the first of Holland's many travels. After this adventurous journey, Bright entered Guy's Hospital as a student in October 1810, where his main teachers included Astley Cooper (p. 11) and Haighton (p. 10) and he was taught medicine by William Babington (1756–1833), whose daughter he married. In 1811 he returned to Edinburgh to study medical subjects and geology and in 1813 he took his degree of MD. He tried being an undergraduate in Cambridge but did not like the life and after two terms at Peterhouse, he took up medical work by becoming assistant physician at the London Dispensary. But soon he was travelling again. He spent about a year travelling in Eastern Europe ending at Waterloo about a fortnight after the battle; he wrote a book about this journey from which it appears that he was not then particularly interested by medical affairs in the cities he visited. In 1816 he was appointed assistant physician at the London Fever Hospital and himself contracted a dangerous illness. He travelled again in 1818 to Western Europe but now with the particular purpose of medical study.

Bright's real working life began in 1820 when he began private practice in London as a physician and was appointed assistant physician to Guy's Hospital. He was elected FRS in 1824 and in the same year succeeded to the position of full physician at Guy's. For a year or so he had lectured in Botany before, in 1825, he become one of the two lecturers in medicine. He continued to give these lectures until he retired in 1843, with Addison his co-lecturer after 1837. He died in 1858, was buried in Kensal Green, and a memorial tablet was placed in St James's, Piccadilly.

Bright was accepted in his lifetime as a great physician. Elected Fellow of the Royal College of Physicians in 1832, he was Goulstonian Lecturer in the following year and later Censor of the College.

He became Physician Extraordinary to the Queen, and was the subject of an article of an article in Pettigrew's *Celebrated Physicians and Surgeons* (1839).

Thomas Addison (1793–1860)

Thomas Addison was a member of a prominent Cumbrian family who for generations lived at Banks House, near Lanercost Priory. Addison is buried in the churchyard of the Priory together with many other members of the Addison family and there is a memorial tablet on the west wall of the Abbey. He was born, however, at Long Benton near Newcastle on Tyne, and was educated at Newcastle Grammar School. Thence in 1812 he went to Edinburgh, qualified MD in 1815 and then went to London to practise as a physician and worked with a great dermatologist, Thomas Bateman (1778–1821). In 1817 Addison was enrolled at Guy's Hospital as a 'perpetual physician's pupil' at a fee of £22 1*s*, became a Licentiate of the Royal College of Physicians in 1819 and Fellow in 1838. After 1817 he never left Guy's, becoming in 1824 assistant physician to Bright and a very successful lecturer in materia medica; in 1837 he was promoted full physician and appointed Lecturer in Medicine, at first in association with Bright. In 1860 he was forced by ill-health to retire, and died soon after in Brighton.

Although in succession to Sir Astley Cooper and Bright, Addison became a third great figure in nineteenth-century Guy's, he was little known outside the Hospital. He was never honoured by the College of Physicians, was not elected Fellow of the Royal Society and on his death there was no obituary notice in the *Lancet*. The most accessible account of his life is by Hale-White (1935) in *Great Doctors of the Nineteenth Century*, apparently compiled from the records of Guy's Hospital; Hale-White was a later physician at Guy's. A contemporary view of his character as a physician was: –

To those who knew him best his power of searching into the complex framework of the body, and dragging the hidden malady to light appeared unrivalled; but we fear that the *one* great object being accomplished, the same energetic power was not devoted to its alleviation or cure. (A collection of the writings of the late Thomas Addison, edited by Wilks and Daldy: New Sydenham Society Publications, 36, 1868).

It was suggested that his lack of interest in treatment might have stemmed from scientific honesty – he knew of no treatment which would be effective.

Addison was a very good lecturer and was worshipped by Guy's students for the quality of his clinical teaching. Yet they feared rather than loved him; he was melancholy, liable to fits of depression, at times appeared haughty and unapproachable, although he was at basis an amiable man. Perhaps his greatest contribution to teaching was his introduction of the practice of students systematically writing reports of their cases, and he had the personality to make this innovation effective. Case reports are the material of his famous book on adrenal function (see below).

Both Bright and Addison added to medical knowledge by the same method. This was firstly to describe and record accurately the symptoms and signs in each case and then, when possible, conduct a careful *post mortem* examination. Bright stated that much could be done for medical science by seeking 'to connect accurate and faithful observations after death with symptoms displayed during life'.

In this way, in 1827 Bright published *Reports of medical cases selected with a view of illustrating the symptoms and cure of diseases by reference to morbid anatomy*. The book was in three parts, of which the first 126 pages of Volume I dealt with 'Disease of the kidney in association with dropsy and albuminuria'. This was the famous description of what came to be known as Bright's disease. Albumin in urine was revealed by holding the specimen in a spoon over the flame of a candle. Bright's disease was further expounded in the Goulstonian Lectures of 1833 and in Guy's Hospital Reports of 1836. One can see a beginning of modern medicine in 1842 when fifty-two beds were set aside for a period of six months for the study of diseases of the kidney, with a small laboratory attached (Cameron, 1954).

Bright also provided accurate reports with *post mortems* in cases of large abdominal tumours, diseases of the pancreas and duodenum, jaundice and disease of the liver.

Similarly Addison described accurately for the first time fatty degeneration of the liver, appendicitis, lobar pneumonia, Addison's anaemia (pernicious anaemia), and what we still call Addison's disease due to degeneration of the adrenals. Addison's anaemia and Addison's disease were both described in a paper given at Guy's Hospital on 15 May 1849, but were not clearly distinguished as two separate diseases. Addison appreciated that disease of the suprarenal capsules might produce general effects and in 1855 published reports of eleven cases where lesions of the suprarenal capsules had been found at post mortem. This was a book of forty pages, illustrated by eleven colour plates to show the colour of the skin and the *post mortem* findings; its title was *On the constitutional and local effects of diseases of the suprarenal capsules*. Not all of the cases would now be accepted as examples of adrenal insufficiency but this book can well be regarded as a beginning of endocrinology. The book was republished in facsimile in 1968 and was 'reviewed' by Keele (1969). When first published in 1855 the book received only a perfunctory review in the *Lancet* but was immediately accepted by Paris physicians; it was Trousseau who in 1856 first used the term 'Maladie d'Addison'.

Physicians 1830–40

In his first principle (p. 18), Hall stated that the first object of physiologists should be to obtain information by observation and he cites instances where, at that time, information might be obtained without experiments. Such sources could be 'the various cases of monstrosity and the interesting facts provided by comparative anatomy. They are a sort of natural experiment ... In addition ... the various results of disease and of accidents afford us examples of undesigned experiments'. Addison provided two examples when he described Addison's disease and Addison's anaemia.

Hall, Bright and Addison were very advanced physicians by the standards of England of 1830–40. At that time the usual physician had a background of general education in the classics, since Fellows of the Royal College of Physicians were almost always graduates of Oxford or Cambridge, but had very little professional training (Newman, 1957). Although they acted as consultants, it seems that they did not usually examine the patient, giving an opinion solely from what they were told. As educated gentlemen they claimed to be the élite of the profession. With the novelists' liberties, Trollope created the richly pompous Dr Filgrave of Barchester (*Dr Thorne*) and George Eliot created Drs Sprague and Minchin of *Middlemarch*.

In Paris, however, there was a new type of physician, represented by R. T. H. Laënnec (1781–1826) and P. C. A. Louis (1787–1872), as described by Singer and Underwood (1962, pp. 172–4 and 721–2). Laënnec is well known as the rather accidental discoverer of the stethoscope, developed into a standard instrument by trial of rolls of paper, tubes made of different woods and of various shapes and even an old oboe (Kervran, 1960). What is more important is that he used auscultation, percussion and palpation to form a routine examination of the chest very like that used today. Laënnec interpreted his physical findings in terms of his knowledge of the pathology of intra-thoracic diseases acquired from his accumulated cases and post mortem examinations. Louis collected his cases, and related clinical findings to post mortem examination, to form description of disease states. He collected numerical data about diseases in the manner of modern statistics; indeed it is said that modern statistical methods can be applied to Louis's data. Marshall Hall knew and several times visited Louis in Paris and Bright also went there; Laënnec was the inspiration of Addison (Keele, 1969). Addison is known to have visited Paris only once, when he was called to attend a member of the Rothschild family. He was then famous in Paris for his description of 'Maladie d'Addison'; the élite of the French medical profession entertained him at a dinner in his honour and Addison replied to the toast of his health in excellent French.

Alongside the physicians of Paris there was also a growing school of physiology, whose work was certainly well known to Hall, but less known to the surgeon-anatomists who were apparently too busy to tour abroad.

Physiology in Paris, 1830–40

At this time in Paris, F. Magendie (1783–1855) was Professor of Physiology and Medicine at the College de France and was carrying out vivisection experiments of an extreme type. The style of his work can be seen from lectures to the College de France published in translation in the Lancet, 1834–5, 1836–7, 1838–9. In these lectures one finds description and argument based on observations as opposed to merely philosophical debate. Throughout the lectures there are passages like, 'These, gentlemen, are mere fictions, the inventions of ingenious and talented men, who were led astray because they neglected to study nature, such as she exists, and preferred creating hypothetical powers, rather than apply the physical conditions of our several membranes and tissues to the explanation of functional phenomena' (*Lancet*, 1834, *27*, 375). The idea of observation and experiment as the method of physiology has always existed and often been ignored. One can find Harvey in 1628 and Dale in 1935 making the same complaint as Magendie in 1834.

Magendie carried out experiments on animals without anaesthesia and he repeated his experiments as demonstrations in his lectures. This made him the target of antivivisectionists, and many aspects of his work were questioned by physiologists such as Bell (p. 6). Sharpey attended one of Magendie's lectures in Paris about 1822 and was disgusted to see him make repeated incisions in the skin of an animal merely to demonstrate that this caused pain – which, as Sharpey observed, we already knew. Similarly, Sharpey saw no value in an experiment in which the stomach of a dog was replaced by a balloon and the animal then given an emetic; Sharpey said that this only demonstrated that a balloon with an open mouth was emptied by compression. In thinking of the character of Magendie, it must be remembered that the facts of life of that time included human surgery without anaesthesia; Magendie, the vivisectionist, was a kind, humane doctor who stayed in Paris attending his patients when his colleagues fled during the plague of 1832.

Marshall Hall, of course, knew Magendie's work. His *Principles of investigation in physiology* cited above are directed against work such as Magendie was doing, sometimes useless because the experiment was not critically designed, sometimes unnecessarily repetitive because the results were already known. One effect of Magendie's work was virtually to preclude animal experimentation in England until the 1870s when anaesthesia could be used. In the intervening years, advances in physiology came from the growing use of adequate microscopes (see Chapter 4).

As a summary to Part I of this book, we may cite Marshall Hall, from his book 'Circulation of the Blood', published in 1831:

What an anomaly it must appear hereafter, that in the early part of the nineteenth century, that department of physic which teaches the nature of the functions of the animal economy, was not recognised as an essential separate branch of the study of physic and surgery in the schools of this metropolis.

References

Bettany, G. T. (1885). *Eminent Doctors: their lives and their work*. 2 vols. London: Hogg.

Cameron, H. C. (1954). *Mr Guy's Hospital 1726–1948*. London: Longmans Green.

Green, J. H. S. (1958). 'Marshall Hall (1790–1857): a biographical study', *Medical History*, 2, 120–33.

Hale-White, W. (1935). *Great Doctors of the Nineteenth Century*. London: Edward Arnold.

Keele, K. D. (1969). 'Addison on the "supra-renal capsules": an essay review', *Medical History*, 13, 195–202.

Kervran, R. (1960). *Laënnec: his life and times* (translated by D. C. Abrahams-Curiel). Oxford: Pergamon.

Munk's Roll of the Royal College of Physicians of London. Vol. II 1701–1800, Vol. III 1801–25. London: the College.

Newman, C. (1957). *The Evolution of Medical Education in the Nineteenth Century*, London: Oxford University Press.

Olmsted, J. M. D. (1944). *François Magendie*. New York: Schuman's (P).

Pettigrew, T. J. (1838–40). *Medical Portrait Gallery: Biographical Memoirs of the most celebrated Physicians, Surgeons, etc.* 4 vols. London: Fisher.

Singer, C. and Underwood, E. A. (1962). *A Short History of Medicine*. Oxford: Clarendon Press.

Thayer, W. S. (1927). 'Richard Bright', *Guy's Hospital Reports*, 77, 253–301 (P).

Biographies

Addison: Hale-White, 106–23.
              Bettany, Vol. 2, 1–14.
              Munk's Roll, III, 205–11.

Alison: *British Medical Journal* (1859), 801–2.
              Dictionary National Biography, **1**, 290–2.
Bright: Hale-White, 63–84.
              Pettigrew, **2** (P).
              Bettany, Vol. 2, 14–23.
              Munk's Roll, III, 155–61.
              Thayer (1927).
Babington: *Munk's Roll*, II, 451–5.
Bateman: Munk's Roll, III, 19–23.
Hall: *Lancet* (1857), ii, 172–5.
              Bettany, Vol. 1, 264–85.
              Green (1958).
              Pettigrew, Vol. 4 (P).
              Memoirs of Marshall Hall, by his widow (1861),
              London: Bentley (P).
Holland: *Lancet* (1873), ii, 650–1.
Laënnec: *Lancet* (1826), **11**, 44–5.
              Kervran (1960).
Louis: *Lancet* (1872), ii, 355–6.
Magendie: *Medical Times and Gazette* (1855), **11**, 558, 583.
              Olmsted (1944).

PART II:
PHYSIOLOGISTS 1835–70

3

Physiology 1835−1870

English textbooks

Whether they were primarily anatomists or later when there were more specialist physiologists, teachers of physiology wrote textbooks, and one way of following the growth in physiological knowledge is to examine the textbooks in their chronological order.

A leading early textbook was *The elements of general anatomy, containing an outline of the organization of the human body*, published in 1829 by R. D. Grainger (p. 8). This book set out the speculations of its time (see p. 12) and there was little in it which resembled or was leading towards modern physiology.

Things had not advanced much in 1840, when the first five of Sharpey's lectures at University College on anatomy and physiology were published in the *Lancet* (Volume 39, 1840−1). For that part of the course called general anatomy, Sharpey (p. 85) recommended Quain's *Anatomy*, which had been begun in 1828 by Jones Quain (p. 72). In editions after 1843, Sharpey himself wrote the first part of the book, about 200 pages entitled General Anatomy; his professorial colleague, Richard Quain (p. 74) wrote the larger part, about 1000 pages on descriptive anatomy, this being similar to the dissecting manuals of later years. Later authors were Schäfer in physiology and Ellis in anatomy.

For his lectures dealing with function, Sharpey recommended his students to read *Elements of physiology* by J. Müller of Berlin. First published in 1832, this was available after 1838 in a translation by W. Baly, a Physician to St Pancras Hospital who had worked with Müller; Sharpey himself had helped Baly with the translated edition. Its accounts of function were only speculative since it used no measurements and referred to no experiments. There was no information about organic chemical substances and chemical formulae were not used.

A similar but smaller book listed by Sharpey was *Human physiology: with which is incorporated much of the elementary part of the Institutiones*

Physiologicae of J. F. Blumenbach by John Elliotson, first published about 1820 and in its fourth edition in 1840.

John Elliotson, FRS (1791–1868)

Elliotson was the son of a chemist and druggist in Southwark. His first medical education was in Edinburgh where he became MD in 1810. In the same year he qualified LRCP (London) and was admitted into Jesus College, Cambridge. He passed MB Cambridge in 1816 and MD in 1821. *Human physiology* was produced while he was a student at Edinburgh and Cambridge.

He settled in London firstly as a physician at St Thomas's Hospital and in 1832 became Professor of Medicine in University College; when University College Hospital was opened in 1834 Elliotson was appointed Senior Physician. He was regarded as a fine teacher and had a large private practice. He had been elected FRCP in 1822, was later Censor of the College and gave several of its formal lectures, notably the Lumleian Lectures of 1829–30 on 'Recent improvements in the art of distinguishing various diseases of the heart': he used the stethoscope. Elliotson had a great desire to be original, leading him into many eccentricities: in 1826 he discarded knee breeches and silk stockings, then the orthodox dress of physicians, and he also wore a beard.

In 1837–8, however, Elliotson became convinced of the reality and therapeutic value of mesmerism and began to use it in his hospital practice and teaching. He was hoodwinked by charlatans and was forced to resign from University College. Thereafter ostracised by the medical profession, he continued to advocate mesmerism, founding a journal, *The Zoist, a journal of cerebral physiology and mesmerism and their application to human welfare.* He finally died in near poverty, forgotten by the medical profession.

Another book recommended by Sharpey in 1840 was *Principles of general and comparative anatomy* by W. B. Carpenter (p. 107). The first edition of this book had then just been published. Carpenter became an important zoologist, remained interested in physiology and was in 1882 elected an honorary member of the Physiological Society. He produced further editions of his book, dividing it into two parts, one specifically devoted to human physiology; finally in 1864 the book was taken over by Power (p. 57).

These books recommended by Sharpey show that in 1840 there was very little that can be related to modern physiology. There was no accurate description of the microscopic structure of tissues. There was no knowledge of organic chemistry. There was no reference to experiments or the use of measurements of functional activity. These are the points we must seek as indicating the beginnings of true physiology.

Real advance is apparent in *The physiological anatomy and physiology of man* by R. B. Todd (p. 36) and W. Bowman (p. 37). Part I was first published in 1843, and Part II in 1856, together with a second edition of Part I. The two volumes of the 1856 edition in the library at Leeds were first owned by T. Clifford Allbutt, 26 Duke Street, St James's, and were presumably bought by Allbutt when he was a student at St George's Hospital in 1858–60. There are some annotations in his handwriting. In this book we find for the first time detailed description of the structure of tissues in the manner of modern histology. Largely, the book reports the work of Bowman using an improved compound microscope (see p. 38) and is illustrated by 298 woodcuts from the author's drawings. The book is of primary importance as marking development from general anatomy to histology but it also refers to experimental work, such as the first attempts to measure arterial blood pressure with a mercury manometer. With the assistance of Beale (p. 42) another edition was published in 1866.

The development of Physiology can be followed through Kirkes's *Handbook of physiology*, first published in 1848. Kirkes was a student at St Bartholomew's Hospital and wrote the book from Paget's lectures (p. 51), with the approval and assistance of Paget. Thereafter the publishers (Walton and later Murray) produced a new edition about every three years, with a succession of authors. Until the 13th edition in 1892 the author was a current lecturer at Barts (p. 53). When after 1892 no author could be found there, the next edition was by Halliburton, Professor of Physiology at King's College, London, and he was succeeded in 1933 by R. J. S. MacDowall.

The early authors were physicians or surgeons, not specialist physiologists, and the avowed purpose of the book was to provide medical students with what they needed to pass their examinations. The 1867 edition which is in the library at Leeds contains simply descriptive and functional anatomy, without reference to experiments; it lags behind the book of Todd and Bowman published in 1856, and is probably little advanced beyond Paget's original lectures. Similarly the 11th edition (1884 by Harris) is far behind Foster's book of 1877. Even in the 1920s–1930s Halliburton was a book liked by many students but not highly recommended by teachers.

Physiology of common life (1859) by G. H. Lewes (p. 118) was written for a general rather than a specialist medical audience; it was first published in serial form in Cornhill Magazine. Although never an examination textbook for medical students, it was widely read and had the effect of interesting many people in physiology; one person so influenced was Pavlov. The book cites experimental evidence mainly from Continental sources such as Bernard and Liebig. It is, however, apparent that the author was a man of literature as well as an experimental scientist. Smith (1960) 'reviewed' the book on the centenary of its publication.

Perhaps the most significant book of the 1860s was *Lessons in elementary physiology*, published in 1866 by T. H. Huxley (p. 112). This was based upon a course of ten lectures given at the School of Mines in 1865, directed particularly to schoolmasters who might teach physiology in schools. Intended as an elementary course, the book gave a concise account of human anatomy, physiological chemistry, histology and experimental physiology more in accordance with modern ideas than any of the preceding books. As an elementary book it did not debate current problems, but stated an outline of physiology as it was collected by Huxley's reading of (mostly Continental) authors. In the first edition, and more particularly in subsequent editions, Huxley acknowledged the advice and help of Michael Foster.

A text-book of physiology by Michael Foster (p. 167), published in 1877 is the first textbook with a modern outlook, although there was obviously much basal physiology still to be discovered. Foster, who was a former student and assistant to Sharpey, acknowledged help he received from Sharpey and Burdon Sanderson. In the hands of Sharpey and his immediate successors, Foster and Burdon Sanderson, physiology was based on sound histology, on knowledge of physiological chemistry, on experiments on man and animals, and on the measurement of physiological functions. The introduction to the book reveals the general advance in scientific education and the status which physiology had attained; Foster said:

I have presupposed my readers to possess a general knowledge of Physics and Chemistry ...
I have further presupposed such an elementary acquaintance with Physiology, as can be gained from Prof. Huxley's *Elementary lessons in physiology* ...
An acquaintance on the part of the reader with both Anatomy and Histology is also taken for granted.

As examples of the modern outlook in the text, we find that in the chapter on respiration, Foster clearly states that O_2 is taken in by cells and CO_2 is produced and that O_2 reaches the cells by diffusion along the gradient of pressure between alveoli and blood and blood and cells; CO_2 similarly diffuses in the opposite direction. In the chapters on the heart, the events in the cardiac cycle are stated and illustrated by recordings of the pressures in the chambers of the heart and great blood vessels. The effect of stimulating the vagus is given. In dealing with nutrition, chemical formulae are used and the formation of urea from protein and the importance of the liver is stated, although there is discussion of the possibility that urea might also be formed from creatine.

Foster's book is well written and one can see why in both Britain and America it became the most influential book in the last twenty years of the nineteenth century. Foster produced seven editions of the book, the last being published in 1897, when the section on the nervous system was largely written by Sherrington.

Physiologists in Europe

In Foster's book of 1877, practically all of the references are to European authors, not British. During the period up to 1870, in France and Germany there were already institutes carrying out physiological research and teaching, but there were none in England. Fitzsimons (1976) has concisely reviewed the activities of nineteenth-century continental physiologists and a useful list with brief notices is also given by Leake (1956).

In the field of physiological chemistry, there were active schools of Wöhler (1800–82) in Göttingen and Liebig (1803–73) and Voit (1831–1908) in Munich; these schools, for example, elucidated the nature of urea and its relation to protein metabolism. In Paris Verdeil and Wurtz (1817–84) were preparing and describing substances such as creatine, creatinine, urea, uric acid, hippuric acid; in 1851, Harley (p. 151), Burdon Sanderson (p. 141) and Marcet (p. 157) went to Paris primarily to study chemistry under these masters, but soon also became students of Claude Bernard.

Leading experimental physiologists were Ludwig (1816–95) in Vienna and Leipzig, and in Paris firstly Magendie (1783–1855) and then Claude Bernard (1813–78), but there were many others. Almost as a random list, Broca (1824–80), Brown-Séquard (1817–94) in France and du Bois Reymond (1818–96), Kollicker (1817–1905), and Helmholtz (1821–94) in Germany are names who were active about 1850 known to physiologists today. Such European physiologists provided most of the evidence in Foster's *Textbook* of 1877 and these were the men who inspired the experimentalists who took over English Physiology after 1870 (see Part III).

Physiology in England

During this period the only physiology in England was that included in the lectures on anatomy and physiology given to students. The medical schools in London and the provinces all had a course of about 140 lectures on anatomy, including physiology and pathology. A course of this length was required to provide the student with one of the certificates needed for entry into the examination for the Licence of the Society of Apothecaries or Membership of the College of Surgeons. There was no requirement as to what was included in the course of lectures. The examination was a single oral examination covering many subjects and it did not require up-to-date physiology.

The lecturers were nearly always surgeons, perhaps with the post of assistant surgeon at a related hospital. Herein lay the difference from Europe. The European physiologists named in the previous paragraph were professional physiologists, with no large clinical commitment and with staff and laboratories provided by the rulers of the state. Research was part of their duties, whereas in England any research was a hobby of the individual and in a laboratory contrived in his own home.

In his opening address to the University of London in 1828 (*Lancet*, 1828–9, *15*, pp. 8–10), Charles Bell saw that advances in science could not be achieved by part-time lecturers and professors:

In the past the temptation of following a lucrative practice, has far outweighed the desire of reputation to be gained by teaching; and, consequently, just when the Professor became useful by the knowledge he was capable of communicating, he has withdrawn himself: and so the situation of a medical teacher, instead of being the highest, and entitling him who holds it to be consulted in cases of difficulty – as being one of the seniors of the profession, one who has withstood petty solicitations, and has maturely studied as well as practised – it is merely looked upon as a situation introductory to business; one of expectancy, and to be occupied in rapid succession by young and inexperienced men.

Let us hope that this University may be able to raise the Professors of science to higher consideration, induce men of talents to prepare themselves for teaching, and to continue their public labours to a later period of life.

He saw hope that the new University might lead to continuity in scientific thought.

The effect of combination into some regular establishment is the uninterrupted progress of science; for, hitherto, those who have taught in our schools, have had successors in their places, without successors to their information or their opinions.

The formal opening of the University would not have been the occasion to point out that the fundamental difficulty was that the professors were not paid enough. The first professors of the University of London were to receive £300 per year for two years; thereafter they would take two-thirds of the fees paid by students attending their lectures. With a class of 200, a very successful lecturer might get £800 per year; in comparison a successful surgeon or physician in practice would probably get £5,000. When in 1851, James Paget (p. 52) decided to resign his positions as Lecturer in Anatomy and Physiology at St Bartholomew's and Warden of the College, his income was £700 per year; a few years later, his income as a surgeon reached £10,000 per year.

With such difference in income, it is not surprising that before 1870 only one man, William Sharpey (p. 78), spent his whole working life as a physiologist. Others, such as Bowman (p. 37) and Paget (p. 49), when aged about 35, retired from physiology to practice surgery.

References

Fitzsimons, J. T. (1976). 'Physiology during the nineteenth century', *Journal of Physiology*, **263**, 16–25P.
Leake, C. D. (1956). *Some Founders of Physiology: Contributors to the Growth of Functional Biology*. Washington: IUPS and American Physiological Society.
Smith, R. E. (1960). 'George Henry Lewes and his "Physiology of Common Life", 1859', *Proceedings of the Royal Society of Medicine*, **53**, 569–74.

Biography

Elliotson: *Medical Times and Gazette* (1868), ii, 164–7.
              Munk's Roll of the Royal College of Physicians, London, Vol. III, 258–62.

Textbooks

Grainger, R. D. (1829). *The Elements of General Anatomy, containing an outline of the organization of the Human Body*. London: Highley.
Quain, R. and Sharpey, W. (1843–8). *Quain's Elements of Anatomy*. 5th edition. London: Taylor.
Müller, J., translated by W. Baly (1838). *Elements of Physiology*. London: Taylor and Walton.
Elliotson, J. (1820). *Human Physiology: with which is incorporated much of the elementary part of the* Institutiones Physiologicae *of J. F. Blumenbach*. London: Longmans.
Carpenter, W. B. (1839). *Principles of General and Comparative Physiology: intended as an introduction to the study of Human Physiology and as a guide to the philosophical pursuit of Natural History*. London: Churchill.
Todd, R. B. and Bowman, W. (1843, 1856). *The Physiological Anatomy and Physiology of Man*. Part I, 1843; Part II, 1856. London: Parker.
Kirkes, W. S. (1848). *Handbook of Physiology*. London: Walton.
Lewes, G. H. (1959–60). *Physiology of Common Life*. London and Edinburgh: MacMillan.
Huxley, T. H. (1866). *Lessons in Elementary Physiology*. London: Macmillan.
Foster, M. (1877). *A Textbook of Physiology*. London: Macmillan.

4

Part-time histologists at King's College 1835 – 70

When the University of London was first founded, a stated purpose was to provide university education with complete religious freedom, in contrast to Oxford and Cambridge which required commitment to the Established Church. Almost immediately King's College, London was established in 1830 by churchmen in opposition to what they called the 'godless foundation' in Gower Street. Later three charters were granted – to the University of London as an examining body able to award degrees and to two colleges of the University – University College, London and King's College. Like University College, King's provided lectures in medical subjects, although it had at first no associated hospital. Particularly there was the usual course in anatomy and physiology and Mayo (p. 7) was appointed Professor. He appointed Richard Partridge (1805–73) as his demonstrator, particularly for the dissecting room.

When, after five troubled years, Mayo resigned in 1835, the decision was made to have two chairs of Anatomy. Partridge was appointed as Professor of Descriptive and Surgical Anatomy, and was also surgeon to the hospital; Todd, the new professor, was responsible for physiology and pathology. This was the same separation which, also in 1836, took Sharpey to University College, London. Partridge remained as Professor of Anatomy until his death in 1873; this chapter deals with his physiological colleagues, Todd, Bowman and Beale. They formed a centre of excellence in histology for the next twenty-five years.

Todd, Bowman and Beale were part-time physiologists; they were also assistant surgeon or physician to King's College Hospital, opened in 1840 largely by the efforts of Todd; links were formed between lecturing posts in the medical school and clinical posts in the hospital. By statute, therefore, the professor of physiology could devote only half his time to the subject and this was taken up by the course of lectures; by 1853 to some 200 regular students and 100 occasional attenders. Despite their excellence as histologists, Bowman and Beale did not have the opportunity to form a research school like the continental institutes.

Robert Bentley Todd, FRS (1809—60)

Todd was born in Dublin, graduated in Arts at Trinity College and obtained his medical qualification at the Richmond School in Dublin. Immediately, in 1831, he went to London, was incorporated at Pembroke College, Oxford, and lectured in the Aldersgate School and the Westminster Hospital. In 1836 Todd was appointed Professor of Physiology and Morbid Anatomy at King's College, London and in the same year became MD (Oxford) and Licentiate of the College of Physicians. As with the appointment of Sharpey at University College in the same year, the intention was to separate the teaching of general anatomy and physiology from the teaching of topographical anatomy; Todd's co-professor at King's College was Partridge. Todd was also appointed physician to King's College Hospital which was then just being established, and indeed Todd was a main figure in its foundation.

In 1835 while at the Westminster Hospital School, Todd began to produce *Cyclopaedia of anatomy and physiology, edited by Robert B. Todd*, published by Longman, Brown, Green, Longmans and Roberts. The first volume of alphbetically arranged articles (A—DEA) appeared in 1836, the fifth and last in 1859. The articles dealt with all aspects of each topic — surgical and topographical anatomy, general anatomy (histology), morbid anatomy, physiology, pathology and a lot of comparative anatomy. Todd commissioned each article from an expert and suffered the usual trials of an editor with his authors. 'A few completely failed to fulfil their engagements, without any assignable reason; others were unavoidably prevented from doing so. In several instances the articles were not completed at the stipulated time'.

In 1840 Todd and Partridge were joined by Bowman and Simon as demonstrators, mostly in the dissecting room; Bowman stayed as lecturer and then co-professor with Todd. Bowman and Todd worked amicably together. Bowman contributed articles to the *Cyclopaedia* and together they wrote *The physiological anatomy and physiology of man* (p.30). The illustrations, particularly of histology, are by Bowman and they both contributed to the text. It is not clear whether Todd actively worked with Bowman in histological investigations; published papers are in Bowman's name alone.

Todd won no permanent place as a physiologist, although in the 1840s he was professor in the most active school of physiology in London. His *Cyclopaedia* was an important book of reference;

his two junior colleagues, Bowman and Simon were carrying out original investigations in histology; Todd and Bowman was a good textbook of physiology. However, Todd himself became heavily involved in the clinical work of King's College Hospital until in 1853 he resigned his chair of Physiology. His successor was L. S. Beale (p. 42). At Todd's resignation some difficulty arose because the post of physician, which Todd wished to retain, was attached to the college professorship and in order to retain his services in the hospital, this rule was rescinded. Perhaps this was a first step towards a full-time professorship in physiology.

Todd was an important physician, described by Collier (*Lancet*, 1934, ii, pp. 855−9) as the greatest clinical neurologist Great Britain had produced until Hughlings Jackson (p. 60). Particularly, Todd recognised the anatomical and physiological basis of disorders then all called 'paraplegia' and he clearly stated the function of the posterior columns and the cerebellum. He can well be regarded as a proper successor to Bell.

Todd's premature death was much regretted and his funeral at Kensal Green was attended by some 200 friends and former students.

Sir William Bowman, FRS (1816−92)

William Bowman was born at Sweetbriar Hall, Hospital Street, Nantwich and was sent to school firstly in Liverpool and then in Birmingham, where his headmaster was Thomas Wright Hill, whose four sons also taught in this advanced sort of school where the students were to govern themselves. One of the sons was Rowland Hill, who in 1840 initiated the penny post and the postage stamp.

William Bowman had three brothers of whom an older brother, Eddowes, became Professor of Classics at Manchester and a younger brother, John Eddowes (1819−55), became Professor of Medical Chemistry at King's College, London. His book, *A practical handbook of medical chemistry* (1853), clearly sets out the medical chemistry taught at that time. John Eddowes, however, died at the young age of thirty-five. At the time of his appointment to King's in 1845, a move to appoint Liebig as Professor of Medical Chemistry was abandoned because Liebig was a Lutheran.

Records of the Bowman family (Thomas, 1966) reveal a family liking and aptitude for drawing. This skill, well developed in William, was important to his research work in histology.

William Bowman's medical education began in Birmingham by

his being apprenticed to Betts, House Surgeon to the Infirmary, and to Joseph Hodgson, a famous surgeon. In Birmingham he also helped a physician, Peyton Blakiston, with measurements of the orifices of the heart. In gratitude Blakiston presented to Bowman a microscope made by Powell (p. 45), the best available microscope of that time. It was with this instrument that Bowman did his original histological work, so the association with Blakiston had important results for physiology.

From Birmingham, Bowman went in 1837 to King's College, London to fulfil the courses necessary for Membership of the Royal College of Surgeons, which he passed in 1839. He was accepted in King's College despite his being a dissenter; the principal of this Church of England College wrote in 1837 'His being a Dissenter will create no difficulty. It is only a question of whether he himself will object to attending Chapel on Sundays'. In 1838 Bowman accompanied a younger man, F. Galton, in travels on the continent, visiting hospitals in Paris, Vienna, Germany and Holland. Galton, who was a cousin of Charles Darwin, became Sir Francis Galton, FRS (p. 123).

After qualification, in 1840 Bowman became Demonstrator in Anatomy and Physiology at King's with R. B. Todd as Professor. Thereafter he and Todd worked easily together. Bowman became lecturer, and finally in 1848, co-professor, with Todd; Bowman retired in 1856 to devote his full time to surgery.

From 1840 onwards Bowman published important original work on histology, using the Powell microscope given to him by his friend in Birmingham. In 1840, in the *Philosophical Transactions of the Royal Society*, there appeared a long paper, with eighty-five drawings by Bowman 'On the minute structure and movements of voluntary muscle'; the paper is in the form of a letter to Todd to be read to the Society. In the following year a further communication on muscle was given by Bowman directly as he had now been elected FRS. A third paper in 1842 is Bowman's best known work – 'On the structure and use of the Malpighian bodies of the kidneys, with observations on the circulation through that gland'. It consists of twenty-five pages with seventeen drawings, some coloured. In 1842 also, there was a short paper in the *Lancet* describing the histology of fatty degeneration of the liver. These are the only formal papers he published on physiology. However during 1840–50 Bowman wrote articles for Todd's *Cyclopaedia* and they were writing *The physiological anatomy and physiology of man*. In each case these articles contained much original work, particularly in Bowman's illustrations in the text book and articles on mucous membranes, muscle, muscular movement, Pacinian bodies in the *Cyclopaedia*.

Bowman always remained a surgeon and was elected FRCS in 1840. He was appointed in 1840 assistant surgeon to the new King's College Hospital and became full surgeon in 1856, at which time he resigned his professorship and ceased teaching Physiology. Although Bowman insisted that he was a general surgeon, he became the greatest ophthalmic surgeon of his time, almost the first such specialist in London, being surgeon to the Royal Ophthalmic Hospital, later Moorfields Eye Hospital. In this field he made many original contributions. He immediately used Helmholtz's ophthalmoscope, appreciated Donders's work on errors of refraction, and prescribed correcting spectacles. He described the histological structure of the cornea and with his friends Donders and von Graefe developed improved operations for cataract and iridectomy for glaucoma. In the treatment of lachrymal obstruction he exchanged ideas with T. Pridgin Teale, Junior, of Leeds. His wide fame and high recognition in his own lifetime was as an ophthalmologist rather than physiologist.

In 1861 Bowman operated on Mrs Gaskell's daughter 'Meta', performing a tonsillectomy. Mrs Gaskell wrote to a friend: 'Meta's tickle operation is to be at half-past one on Monday. I have a note from Mr Bowman recommending that Miss Gaskell should have a glass of sherry at one o'clock on Monday. I think Mrs. Gaskell will have one too'.

Many of Bowman's personal letters survive (Thomas, 1966) to show his family relationship and also his contacts with great people of the mid-Victorian era.

There are letters from Florence Nightingale, one dated from Scutari on 14 November 1854, describing conditions there shortly after the Battle of Balaclava on 25 October. Later letters from Florence Nightingale refer to the training of nurses at King's College Hospital. She had met Bowman in 1850 when training as a nurse at the Harley Street Institution for the Care of Sick Gentlewomen, where he was surgeon. They had also been associated in the foundation of St John's House, a training school for nurses on the religious basis that the work of nursing is a holy function; later this institution provided the nurses for King's College Hospital.

There is also a series of ten letters from Charles Darwin written between 1866 and 1878 and others from F.C. Donders (1818–89), a Dutchman whom Bowman had met at the Great Exhibition, 1851. Much of this correspondence involves questions that Darwin had asked about the effects of emotion in producing tears and contraction of the *orbicularis oculi* muscles; these questions arose from a book that Darwin was writing on *The expression of emotions in men and animals* (1872). In these letters, it is revealed that Donders sent Darwin a small book in Dutch, written in 1847, which contains many of the ideas Darwin published in *The origin of species* (1859). In 1869, Bowman took Donders to Downe to meet Sir Charles Darwin.

However, to return to Bowman, the physiologist and famous ophthalmic surgeon, knighted in 1884; in 1889, some 600 subscribers united to provide tribute to Bowman, then still living. The tribute was his portrait by W. W. Ouless and the republication of his collected papers in two volumes. The first volume, entitled *Researches in physiological anatomy* was edited by Burdon Sanderson. The second volume contained his papers on ophthalmology. The book was published in 1892; the editors had the co-operation of Bowman himself but he died shortly before its publication. Earlier, in 1882, physiologists had shown their appreciation by electing him an honorary member of the Physiological Society; Paget was similarly honoured at the same time. Bowman was invited to be a founder member of the Society but in 1876 he felt that he was too far removed from the science to be an active member.

Sir John Simon, FRS (1816–1904)
John Simon was born in London, one of a family of fourteen. Both parents were half French; his father became a prosperous member of the London Stock Exchange. After schooling at Burney's school in Greenwich and a period in Germany mastering European languages, John began his medical education in 1833 by being apprenticed at a fee of 500 guineas to J.H. Green, surgeon at St Thomas's Hospital and newly appointed Professor of Surgery at King's College, London. Simon attended lectures at both institutions, doing his anatomy and dissections at King's College, and in 1837–8 he was prosector for Todd's lectures. He passed Member of the Royal College of Surgeons in 1838 and was appointed joint demonstrator in anatomy at King's, his colleague being Bowman. In 1840 when King's College Hospital was opened, Simon was appointed Senior Assistant Surgeon and Bowman was his junior.

However he had not much to do in his professional appointment and he used this waiting time to read extensively in Oriental languages, steep himself in English literature and philosophy and formed friendships with many literary figures of the time. He was introduced into this circle by his master, J.H. Green, who was an intimate of Coleridge and became Coleridge's literary executor. Amongst the friends Simon made at this time was G.H. Lewes (p. 118).

During this waiting period, Simon's interest was aroused by the microscopic work of his colleague Bowman and this led to some scientific papers. Simon wrote the article on 'Neck' for Todd's *Cyclopaedia* (1842) and essays on the Thymus Gland (1844) and Thyroid (1844). These were well received and he was elected FRS in 1845. They are mostly comparative anatomy and histology. In 1847 he published a paper on 'Subacute Inflammation of the Kidneys', describing histological appearances and showing the trend of his interests towards pathology.

Simon's life work began in 1847–8 by three appointments. Firstly he was appointed lecturer in pathology at St Thomas's, the first specific appointment in pathology in this country. Secondly, he was appointed Assistant Surgeon at St Thomas's. Neither of these positions carried much money and presumably it was the salary of £500 per year which led him also to accept appointment as Medical Officer of Health for the City of London, the only other such appointment at this time being in Liverpool. For the next twenty-five years there were two sides to his career. On the one hand he was pathologist and surgeon at St Thomas's Hospital. His lectures took up new ideas as they came; as a surgeon he was advanced enough to boil instruments and wash hands in the years preceding the Listerian revolution. On the other hand, he was active as Medical Officer of Health, visiting, inspecting and writing brilliant reports on trouble points and on outbreaks of infectious disease. He introduced many sanitary reforms which were needed to meet epidemics or endemic states of cholera, typhus, typhoid, diphtheria, anthrax, smallpox and tuberculosis. This work was continued on a wider scale after 1855 when he was appointed Medical Officer to the General Board of Health, and thereafter until his retirement in 1876 he was the chief adviser to governments on medical matters under a series of titles. He acquired enormous acceptance in this work and from 1870 to 1890 he and James Paget were grand men of medical and literary life in London. Simon was made KCB in 1887, the Queen's Jubilee year.

Apart from his early period at King's, Simon did no original experimental work, but he influenced physiology in several ways. Burdon Sanderson, appointed Medical Officer of Health to Paddington in 1855, carried out his duties in the way that had been used by Simon in the City of London. With similar outlooks on 'infectious' diseases, Simon encouraged and supported Burdon Sanderson in investigations on the transmission of such diseases, which formed the first phase of Burdon Sanderson's career in experimental work (p. 141).

Simon was particularly interested in medical education and in his capacity of Medical Adviser to the Privy Council prepared the draft of the Medical Act of 1858, setting up the General Medical Council; in fact the act as passed was an emasculated version of the much more precise act that Simon had drafted. Thereafter Simon was a vociferous critic of the Act and used all means of private persuasion and public outcry to produce a strengthening of it. His position as a government official prevented his actually being a member of the GMC until after his retirement. Nominated as a Crown Member in 1876, he became the driving force of the GMC for the next ten years, leading to the Royal Commission of 1881 and the new Act of 1886. Simon's great purpose was to have only one qualification for the whole country, rather than perpetuating the system by which individual institutions and universities held their own examinations and issued their own licence to practise.

From his established position of scientific authority, Simon was a strong anti-antivivisectionist, opposing the Act of 1876 and in 1881 in his presidential address to the section of Public Health at the International Medical Congress of 1881, Simon spoke strongly about the way in which prejudice and the working of the Act was hindering scientific research into infectious diseases.

Lionel Smith Beale, FRS (1828–1906)

Beale, the son of a surgeon and officer of health in London, was born near St Paul's, Covent Garden, and in 1841, aged only thirteen, was apprenticed to a surgeon in Islington. He was a brilliant student, who matriculated in the University of London in 1847, entering King's College; he passed Licentiate of the Society of Apothecaries in 1849 and MB London in 1851. During his studentship, he spent time as anatomical assistant to Professor Acland at Oxford. After qualifying he became resident physician at King's College Hospital. Having learnt microscopic technique from Bowman, he set up a private laboratory where he taught students chemistry and microscopy. This did not last long for, in 1853, when Todd retired, he was elected Professor of Physiology and Morbid Anatomy. He was preferred to Huxley, possibly because he was a churchman and Huxley was not.

For the next twelve years Beale lectured at King's on physiology and morbid anatomy; lectures in descriptive anatomy were still given by Partridge. Between 1851 and 1872 he was very active in research with the microscope, publishing numerous papers on subjects such as the anatomy of the biliary ducts, nerve endings in voluntary muscle, sarcolemma, branching of nerves in the bladder, terminations of nerve fibres, dentine and nerves to capillaries. He also described amoeboid movements. Following some of the earlier papers he was elected FRS in 1857.

Like Bowman he had the artistic skill to illustrate his papers by his own elegant drawings. His work progressed beyond Bowman's because he fixed the tissues and he used stains, carmine and aniline dyes, allowing better definition of structures. Because they stained with carmine he described the nucleus and protoplasm of cells as 'germinal matter' in comparision with extracellular material which he called 'formed matter'. In his views (see below) it was the germinal matter which was endowed with vitality, whereas the formed matter merely followed physical laws.

Beale wrote two large books on the techniques of the microscope — *The Microscope in clinical medicine* (1854) and *How to work with the microscope* (1857). The first deals with the use of the microscope in medical diagnosis including pathology; the second was written for the general use of members of the Society for Microscopy of which Beale was an active member and ultimately President. Both books were in sufficient demand to require four or five editions before 1880; they were important advanced manuals. There was also in 1866 a new edition of *The physiological anatomy and physiology of man* (see p. 30) with the authors now Todd, Bowman and Beale.

On clinical topics his books were *Urinary and renal derangements and calculus disorders*, *Slight ailments*, and *The liver*, these also being rewritten in second editions. Though all this activity as microscopist, teacher and writer, he continued to work as a physician. He finally resigned the post of Professor of Physiology in 1869, being succeeded by Rutherford. However he continued as pathologist and physician at King's College Hospital and was appointed Professor of Medicine in 1876, retiring in 1896 because of a cerebral thrombosis.

In his work in histology and in medicine, Beale was a scientist of his time. He was also a man of strong religious convictions. In a book, *Life theories: their influence on religious thought* (1871) he set out this dual approach, reaching the position that the 'physical theory of life' cannot be accepted and there must be a 'theory of vitality'. He did not accept Darwinism. Something of these ideas appears in the first ten lectures of his course to students at King's in 1864, which were published under the title *On the structure and growth of tissues and on life*. Biographers of the time indicate that these ideas obscured the value of his scientific work; ideas such as 'germinal matter' versus 'formed matter' tended to obscure his histological descriptions. Nevertheless many of the next generation found inspiration in his books, as testified by William Osler in an obituary notice in the *Lancet* (1906, i, pp. 1004–7).

The microscope 1835–70

The first advance of physiology from vagueness to precision was because of the compound miscroscope. Lecture courses in anatomy and physiology prior to 1840 began with what was called 'general anatomy'. The simple microscope had been insufficient to define the histological structure of tissues. General anatomy as taught, for example, by Sharpey in 1840 and as set out in books of that time (see p. 28) started with the ideas of Bichat, who about 1800 had suggested that the tissues were composed of a few common 'textures' but the simple microscopes then available might distinguish but could not clearly define cells. The compound microscope, available in 1840 to Bowman, could distinguish cells and so precise histology could replace the old 'general anatomy'. Similarly pathology, as we know it, began with the compound microscope.

Of physiologists working with the old simple microscope we may cite Sharpey (p. 82), who in Edinburgh in 1835 observed the movement of fluid caused by cilia in tadpoles and mytilus: 'for the most part I used a doublet lens of 1/35th inch focus'. We may also cite the experience of James Paget (p. 52). His first work in pathology was the recognition of the worm *Trichina Spiralis* in spots in affected muscles. This was in 1834 when he was still a student. He and others had observed the small spots and Paget wished to examine them under the microscope, but there was no microscope at St Bartholomew's Hospital. He was introduced to the Head of the Natural Science Department of the British Museum but he had no microscope. Next he was sent to the Head of the Department of Botany, who had a single lens microscope adequate to show the curled-up worm.

The rise of histology in the 1840s was the natural result of the development of the compound microscope into an instrument with resolution about equal to the ordinary student microscope of today. Important developments in the theory and practice of achromatic lens were made in 1830 by J. J. Lister (1786—1869), father of the surgeon, Lord Lister. J. J. Lister encouraged the making of microscopes to use these lenses and good compound microscopes were made in London in three workshops, those of Ross, Powell and Smith. By 1839 microscopes were sufficiently widely available for the formation in London of the Microscopical Society which soon had about 100 members, partly scientists and partly wealthy amateurs. The Society encouraged Ross, Powell and Smith to produce better microscopes of good mechanical design and workmanship. The microscope of Powell and Lealand after 1840 cost about £10, had coarse and fine adjustments, a substage condensor and was capable of defining an object of 1 micron. These were research instruments and in 1840—50 the English microscopes of Powell, Ross and Smith were at least as good as any available on the continent. The English manufacturers were also encouraged to produce simpler and cheaper microscopes, costing about £5, adequate for teaching students. However the greater quantity production in Germany and France meant that student microscopes from the continent were cheaper and these were what was mostly used in English medical schools after 1860.

The limitations of the work of Bowman and other microscopists of his time were not the resolving power of their instruments but the lack of effective preparation of the tissues. Tissues were teased apart to reveal individual cells and structures or were merely examined from the surface or by light transmitted through thick sections. Beale stained tissues with carmine and sometimes used a process of fixation but the cutting of thin sections was not available until 1870—80, when microtomes were used and the oil-immersion objective also became available.

Teachers such as Sharpey at University College, Quekett at the College of Surgeons and Acland at Oxford wanted their students to see miscroscopic structure for themselves. When microscopes were still somewhat rare, systems were installed by which a single microscope with its light source could be circulated along a rail track to students seated at a table or lecture theatre bench; Sharpey's circular table with rail track for the microscope still exists in the Librarian's Office at University College.

Finally in 1857, histology was sufficiently developed and the microscopes available for Sharpey to introduce at University College practical classes in histology. In addition to Sharpey's five lectures a week on anatomy and physiology, the *Lancet* of 1857 announces three classes a week on practical histology. These were to be given by G. Harley and were not compulsory for the degree of the University of London (see p. 86). At almost the same time, Hughes Bennett in Edinburgh instituted classes in microscopy (p. 96). In other schools there might be one class per week of Microscopical

Demonstrations. As histology developed and microscopes became available, histology, with the students making their own preparations, had become one component of their practical classes by about 1870. It is recorded that in 1871, the Leeds Medical School spent £60 on microscopes. In 1884 the advertisement for the first Professor of Physiology in Leeds stated: 'A few microscopes are kept for demonstrations but each student is expected to provide one for himself under the direction of the Professor'.

Within the general field of medical science, the microscope created two new and precise disciplines. After 1840, the old 'general anatomy' was replaced by 'histology' and the papers of Bowman published in 1840–2 are outstanding examples of the new science of exact microscopic anatomy and the reasoning which can be derived from it. By 1877, histology was so recognised as a separate subject that it was not necessarily regarded as part of physiology and Foster could say that he expected readers of his textbook of physiology to have a knowledge of anatomy and histology (p. 31).

A second new discipline was 'pathology'. The work of Bright and Addison (p. 21) had introduced to England precise description of gross morbid anatomy. Simon and Paget (p. 49) used the new microscopes to give precise description of minute morbid anatomy, and so became the first English pathologists, lecturing specifically on this subject. Beale can also be regarded as a founder of English pathology (Foster, 1958).

Up to 1870 the microscope was the new technique which turned physiology into an exact descriptive science; after 1870 it was largely displaced in this role by experiments and accurate measurement of function (Part III). It is said that Beale resigned as Professor of Physiology bcause he did not wish to devote time to learning the new techniques of experimentation and he was succeeded in 1869 by an experimenter, Rutherford (p. 187).

References

Bradbury, S. (1967). *The evolution of the microscope*. Oxford: Pergamon.

Foster, W. D. (1958). 'Lionel Smith Beale (1828–1906) and the beginnings of clinical pathology', *Medical History*, **2**, 269–73.

Hearnshaw, F. J. C. (1929). *The Centenary History of King's College, London*. London: Harrap.

Lambert, R. (1963). *Sir John Simon 1816–1904 and English Social Administration*. London: MacGibbon and Kee (P).

Thomas, K. B. (1966). 'The Manuscripts of Sir William Bowman', *Medical History*, **10**, 245–56 (P).

Biographies

Beale: *Proc. Roy. Soc. London B* (1906–7), **79**, lvii–lxiii.
              Lancet (1906), i, 1004–7 (P).
              Foster (1958).

Bowman: *Proc. Roy. Soc. London* (1892–3), **52**, i–vii.
             British Medical Journal (1892), i, 742–5 (P).
             Lancet (1892), i, 779–81.
             Collected papers of Sir W. Bowman, ed. J. Burdon Sanderson
             and J.W. Hulke (2 vols, 1892) (P). London: Harrison.
             Thomas (1966).
J.J. Lister: *Dictionary of National Biography*, 33, 347–50 (by his son, J.L.).
Partridge: *Medical Times and Gazett* (1873), i, 347–8.
Simon: *Lancet* (1904), ii, 320–5 (P).
             British Medical Journal (1904), ii, 265–7 (P).
             Lambert (1963).
Todd: *Lancet* (1860), i, 151.
             Munk's Roll of the Royal College of Physicians of London.
             Vol. IV, 15–16.

5

Part-time teachers in medical schools

Teaching in the hospitals

After 1830 the great London Hospitals began to accept more responsibility for their students. Instead of allowing students, for a fee, to walk the wards and watch operations, the hospitals began to provide organised courses of lectures in competition with the private medical schools which they were replacing. After 1835 more and more of the hospitals formed colleges, in which the better students registered to follow a regular course of instruction towards the qualifying examinations of the Royal College of Surgeons or the Society of Apothecaries, and, for a few, the Degrees of the University of London. Registration was made compulsory after 1858, when students were required to register with the General Medical Council, stating their college. The Medical Registration Act of 1858 made compulsory what was already becoming accepted practice (Newman, 1957).

There had to be courses of lectures under the title 'anatomy, physiology and pathology' to meet the requirements of the College of Surgeons, the Society of Apothecaries, and later the University of London. Within the period of this chapter there was no requirement for practical classes in physiology. The courses listed in the *Lancet* in September each year indicate increasing separation of lectures in general anatomy, physiology and pathology from the anatomy of surgery and the dissecting room.

The lecturers were always young clinicians, more often surgeons than physicians; none of the hospitals created full-time lectureships in physiology until after 1880. Similarly the lecturers in the provincial medical schools were clinicians. Paget, Savory and Baker at St Bartholomew's, Jones at Charing Cross and De Morgan at the Middlesex were typical examples of the teacher of anatomy and physiology in the middle years of the nineteenth century. Primarily he would be a surgeon; usually being appointed demonstrator in Anatomy soon after qualification and then becoming assistant surgeon and lecturer in anatomy and physiology. He would give these lectures for ten to fifteen years, until appointment as full surgeon, and then he would often change to become lecturer in surgery. In the earlier years of his career, the lecturing and demonstrating gave a surgeon some income before he had acquired private practice and, in addition, young surgeons become known and able to gain private practice in this way.

No provision was made for lecturers to indulge in research work, and most of them made no contribution to the advance of physiology. Those who did, did so by work in a private laboratory. As clinicians, any time abroad was spent under great clinicians in Europe, so that most lecturers had no experience of experimental physiology as it was then being carried on in Europe. The system of part-time lecturers was criticised by Bell in his opening address to the University of London (p. 33), but it continued to be the usual arrangement in medical schools until after 1870. For physiology, it was a bad system which prevented the development of schools of physiology in this country.

The following biographies are of Englishmen who made a contribution to physiology either as prominent lecturers, writers of textbooks, or by experimental work, mostly with the microscope. Most of the lecturers of about 1850 were still active clinicians when the Physiological Society was formed in 1876: a man can be accepted as a physiologist if he is recorded in the early minutes of the Society as a founder member or as being present as a guest at early meetings (Sharpey-Schafer, 1927).

St Bartholomew's Hospital Medical College

The story of the hospital is related by Moore (1918) but the foundation of the medical school can best be told through the life of Sir James Paget. He was a very great figure in the foundation of the school and has left an autobiographical account of his life from the time he entered the hospital as a student in 1834.

Sir James Paget, FRS (1814–99)

James Paget was the youngest son of a merchant at Yarmouth. His older brothers were sent to Charterhouse and one of them, Sir George Paget (p. 164) obtained a medical fellowship to Caius College and ultimately in 1872 became Regius Professor of Physic at Cambridge. When James's turn came, the family fortunes had declined and there was no money; instead he went to the local school and from 1830–4 was apprenticed to a surgeon in Yarmouth for a premium of 100 guineas. In 1834–6 he was a student at St Bartholomew's Hospital, living very frugally, and in May 1836 he passed the examination for Membership of the College of Surgeons. James Paget thus followed the medical education available in about 1830 to men of limited means. In an autobiographical memoir written fifty years later (*Memoirs and letters of Sir James Paget*, edited by Stephen Paget, one of his sons), James Paget discusses his experiences of apprenticeship and as a student in London in 1830–6.

An apprentice could merely be a dispenser for his master and acquire very little medical training. Paget's master helped him, taught him some medicine and trained him in the accurate keeping of notes of cases. An apprentice had much unoccupied time, and what the apprentice gained depended largely on how he used this spare time. Paget used his time in medical reading, learnt French and made a botanical survey of the district. With his brother Charles, he published in 1834 *A sketch of the natural history of Yarmouth and its neighbourhood, containing catalogues of the species of animals, birds, reptiles, fish, insects and plants at present known* by C. J. and James Paget. The book includes 88 closely printed pages of catalogue giving the names of 766 insects, 729 flowering plants and 456 non-flowering plants. The introduction and all but the entymology was by James and is illustrated by his drawings – a remarkable book for a man of only twenty. Paget thus began his hospital student period with some knowledge of anatomy and physiology and some experience of medicine and surgery, together with an unusually educated power of observing and recording.

From Paget's account it is apparent that the system which required students to spend two years in London acquiring the certificates required by the Society of Apothecaries or College of Surgeons provided at one extreme the possibility of idleness and worse (Bob Sawyer and Ben Allen of *Pickwick Papers*). At the other extreme was Paget, who used the opportunity to read and study and become as well educated professionally as was then possible.

In his memoir of 1880, Paget recalls the books he read in his hospital years. In Physiology his main reading was Müller's *Textbook of physiology*. This was not yet available in translation (see p. 28) and Paget taught himself German in order to read it. He also records that he read lectures in the *Lancet* and found them most useful, an interesting independent justification of Wakley's purpose in publishing them. Paget's ability to read German brought him into contact with Marshall Hall (p. 15), for whom he did some translation work. Although he became quite a friend of the older man, Paget said that he found Marshall Hall apt to be rather tedious because of his repetition of his ideas about reflex action.

When he passed MRCS Paget was recognised by staff and his fellow students as a brilliant student and a reliable, very attractive personality, but this could not gain him a hospital appointment before the hospital apprentices. From 1836–43 Paget had no major hospital appointment and virtually no private surgical practice. He made a scanty income as subeditor for medical periodicals such as *Medical gazette and quarterly journal*, he was curator of the museum at St Bartholomew's Hospital and classified the collection in the Museum of the College of Surgeons.

In 1843 teaching at Barts was reorganised with the decision to separate lectures on general anatomy and physiology from those on descriptive anatomy, as had been done at University College and King's College some seven years earlier. Paget was appointed Lecturer in General Anatomy and Physiology. Also a college was set up to provide supervision of all students at Barts and residence for about twenty-five students. Paget was appointed Warden and lived in the College which was contrived from six houses in Duke Street. In 1847 he had a further appointment as Professor of Anatomy and Surgery at the College of Surgeons, where his lectures achieved wide acclaim .

So for eight years from 1843–51 Paget gave some 150 lectures a year on Physiology – lectures which, although apparently *ex tempore*, were carefully prepared and for which he repeated microscopic and experimental preparations to produce illustrative diagrams. These lectures, with his collaboration, were made into a book by Kirkes, one of this pupils. This book went through repeated editions (see p. 30) into modern times. He was thoroughly versed in the physiological literature of that time, being able to read both German and French; he commented that it was then possible for one man to keep up with the literature of the whole subject. Although he made virtually no original contribution, he was recognised as a great teacher. Michael Foster later wrote that at the time of Paget's lectures, he and Sharpey were 'the only physiologists devoting themselves entirely to this subject'. Others were surgeons who gave lectures. His mastery of the literature is illustrated by a Croonian Lecture (1857) 'On the cause of the rhythmic motion of the heart' in which there are indications that he had himself verified by experiment the statements of others. He also published a paper in 1850 observing and studying the fact that a living egg can be cooled many degrees below 32°F without freezing. He was elected FRS in 1851. The general admiration of Paget's early period as a physiologist was expressed in 1882 by his election as an Honorary Member of the Physiological Society.

His income from his work as lecturer in Physiology and Warden of the College was £700 a year, barely enough for comfortable living. By 1851 he had acquired sufficient reputation amongst staff at Barts to allow him to resign his lectureship and wardenship and enter private practice; also he says he was beginning to find students tedious. He was progressively more successful, was appointed surgeon to Barts, and built up an enormous practice yielding as much as £10,000 per year. From about 1858 Paget was the leading surgeon in London, surgeon to the Royal family and an almost automatic choice as a consultant. He also became an eminent figure – chairman of committees such as those on rabies and vaccination; President of the Royal College of Surgeons, of the Pathological Society and of the International Congress of Medicine held in London in 1881; Vice-chancellor of the University of London, member of the General Medical Council and so on. A baronetcy was conferred on him in 1871. He was a superb lecturer and gave many formal addresses; it must have been a considerable capture when he gave the opening lecture in the new building of the Leeds Medical School in Park Street on 3 October 1865. His house in Harewood Place, Hanover Square was a great social centre where he was at home to his many friends, including Tennyson, Browning and George Eliot, Sharpey, Marshall Hall and Bowman, Huxley, Darwin and many others – indeed practically all the great figures of Victorian London.

Paget's great scientific work lay in the foundation of Pathology and he is linked with Virchow (1821–1902), of whom he was a friend and admirer. His name is still attached to Paget's Disease of the Nipple and of Bone. His first pathological work was the recognition of the worm *Trichina Spiralis* in spots in affected muscle. This was in 1834 when he was still a student, and the story of this episode is related on p. 44. Simon (p. 40) was a great contemporary pathologist.

Paget's colleagues and successors at St Bartholomew's

William Senhouse Kirkes (1823–64)
The name of Kirkes is known as the author of the *Handbook of Physiology* (see p. 30) first published in 1848. Kirkes, however, was a physician and never actually lectured in anatomy and physiology. He was a student at Barts from 1841–6; he wrote the first edition of the book from his and Paget's notes of Paget's lectures. He was elected FRCP in 1855. In the hospital he was appointed assistant

physician in 1854 and physician in 1864 but died in that same year. He published an important paper on *Embolism*.

Sir William Scovell Savory, FRS (1826–95)

Paget's immediate successor was Savory, who subsequently became a prominent surgeon. Savory was born in London and was a student at Barts from 1844–7, qualifying Member of the Royal College of Surgeons in 1847 and MB, London in 1848; he was elected FRCS in 1852.

In 1849–59 he was demonstrator in anatomy under Paget and was also Tutor to students taking the University of London degree. During this period he produced papers on the valves of the heart and on the development of striated muscle fibres in mammalia and was elected FRS in 1858. In 1859 he succeeded Paget as Lecturer in General Anatomy and Physiology and gave these lectures for ten years; for three years he also lectured on comparative anatomy and physiology at the College of Surgeons. Following Kirkes, Savory produced the fourth and fifth editions of Kirkes's *Handbook*.

His career was in surgery, however. He had been appointed assistant surgeon at Barts in 1861 and full surgeon in 1867. In 1869 he became lecturer in surgery and gave up his lecturership in anatomy and physiology. Thereafter, he was a powerful figure at St Bartholomew's, member of the Council of the College of Surgeons, and President from 1885–9, sitting on many committees and commissions. He was appointed surgeon extraordinary to Queen Victoria, and was made a baronet. Savory remained an old-fashioned surgeon who, even after 1880, still opposed the methods of Lister.

William Morrant Baker (1839–96)

Baker succeeded Savory both as Lecturer in Anatomy and Physiology and as author of Kirkes's *Handbook*. He had been a student from 1858–61, thus attending lectures by Paget. He qualified MRCS in 1861 and became a Fellow of the College of Surgeons in 1864. From 1864 he was Demonstrator in Anatomy and in 1869 succeeded Savory as Lecturer in General Anatomy and Physiology. Baker continued to give these lectures until 1885 and produced six editions of Kirkes's *Handbook*. From 1867–73 he was the Warden, resident in the College and was also Secretary of the Medical School; during his period as Secretary the annual intake of the school reached 100 for the first time.

Baker was appointed assistant surgeon in 1871, and then surgeon in 1882 until he resigned in 1892.

Charing Cross Medical School

Thomas Wharton Jones, FRS (1808–91)
When Thomas Wharton Jones was born, his father was Secretary to the Customs at St Andrews; both of his parents came from families of high standing, but none of this affluence came to Thomas Wharton. He was educated at schools in Stirling, Dalmeny and Musselburgh, before entering literary classes in the University of Edinburgh in 1822. In 1825 he began to study medicine, qualifying Licentiate of the Royal College of Surgeons, Edinburgh in 1827.

Immediately he became assistant to the great teacher of Anatomy, Robert Knox, and was involved in the Burke and Hare affair as one of the anatomists who used the murdered bodies. Actually he was little involved but the scandal drove him out of Edinburgh to Glasgow, where he became assistant to the ophthalmic surgeon, McKenzie. In 1835 he began practice in Cork and in 1837 he visited centres in Europe before settling in London in 1838. He passed the examination for Membership of the College of Surgeons in 1841 and was elected FRCS in 1844.

In London Jones practised as an oculist and, from work he had done in Glasgow, was known also as a physiologist and anatomist. He was appointed Lecturer in Physiology at Charing Cross Hospital, where he lectured from 1841 to 1850; Huxley was a pupil in whom Jones inspired an interest in Physiology. While in Glasgow he had already made important observations with the microscope and he continued this work at his house in George Street, Hanover Square. His work recognised by election as FRS in 1840.

Finally he was appointed Ophthalmic Surgeon at University College Hospital and Professor of Ophthalmology at University College in 1851, which post he continued to fill until his retirement from ill-health in 1881. He retired to live at Ventnor where he died in 1891.

Although Jones lectured in physiology for only ten years, he worked in physiology throughout his life and, particularly before 1850, he contributed important original observations. We have the testimony of Huxley that he was a good lecturer, inspiring because of his lucid, logical and coldly scientific presentation. Indeed, as Huxley developed his own prowess as a lecturer, a model could have been his first teacher, Wharton Jones. In 1835, while in Glasgow, Jones described the ovum and the changes which occur after fertilization, including the formation of the chorion. In 1846, while he was at Charing Cross, he described the amoeboid movements of leucocytes and also described the microscopic appearances in early inflammation in tissues such as the frog mesentery and bat's wing, easily available to microscopic observation. These observations were forerunners to Lister's more detailed description of inflammation in 1857; for two years after Jones had become ophthalmic surgeon at University College Hospital, Lister was his assistant and carried Jones's methods forward into his own classical investigations. For his work Jones was awarded the Astley Cooper Prize and his results were published in the Guy's Hospital Reports of 1850. In this paper he also reported changes in a nerve beyond a site of section and it seems that A. V. Waller (p. 64) took up this topic on the suggestion of Jones.

Jones apparently regarded his work on inflammation as his great contribution and he reverted to it throughout his life. Essentially the same work was republished as an appendix to a book on *Failure of sight from railway and other injuries of the spine and head* (1869) and also in 1891 shortly before his death as a separate book *Report on the state of the blood and the blood vessels in inflammation; and on other points relating to the circulation in the extreme vessels; with report on lymphatic hearts, and on the propulsion of lymph from them through a proper duct into their respective veins.*

Jones was a much respected ophthalmologist and wrote a treatise *Ophthalmic medicine and surgery*, which was reprinted several times, and he also wrote *Diseases of the ear*. He carried out well the work of ophthalmology in University College; his lectures were not a required part of the course for any of the examinations and his main influence was through a series of assistants including Lister, Tweedy (his successor at University College Hospital) and Mott, who later became Lecturer in Physiology at Charing Cross. Jones was the first specialist ophthalmologist at University College Hospital.

Jones was a man of strong religious convictions and was interested in general biology, writing essays such as *The wisdom and benefice of the almighty, displayed in the sense of vision* (1851). He did not accept Darwinism. Two lectures at University College in 1874 and 1875 were entitled 'The Evolution of the human race from apes, and of apes from the lower animals, a doctrine unsanctioned by science' and included the statement that 'Natural Selection in the sense in which it is applied to Evolution by Mr Darwin is a mere conceit'.

From the above outline of his life, Jones should have been an accepted figure amongst physiologists and amongst clinicians at University College Hospital. In fact he did not marry and made no friends in either field of his work. He felt himself spurned, beginning with the Burke and Hare episode, where he was essentially innocent but was subjected to public outcry and received no sympathetic help from Sharpey then, or later when Knox and Jones had both come to London. In his later scientific work Jones mistrusted the improved microscopes, and the use of thin sections and staining which were introduced after his early work had been done, regarding much of what they showed as being artefacts. He was aggressive against the newer generation of scientists, lamenting publicly and privately the sins of what he called the Royal Society clique. Somewhat naturally, they rejected him and he retired into an impenetrable shell. In the last year of his life the republication of his work on inflammation contained criticism of Burdon Sanderson, M'Kendrick, Stirling and Lister, as not following correct inductive reasoning. Huxley remained loyal to his old teacher, saying in an Obituary note in the British medical journal (1891, ii, p. 1177) that he felt that in fact Jones had a great deal to complain of in the way he had been ignored by scientists. But in writing of a visit to Ventnor shortly before Jones's death, Huxley reveals the difficult nature of the man: 'I had to listen to some rather sharp tirades against modern physiology in general, and certain excellent friends of mine in particular; indeed I am not sure but that there were some glances at my own doctrines and misdeeds'.

Thus, when in 1881 a letter arrived at University College Hospital in which Jones said he could not fulfil his duties because of illness, there were no friends and no one knew anything about his private life. His assistant Tweedy sought him out, and found him penniless and starving in an unheated room in the 1881 blizzard. A collection, including from Ringer and Erichsen, produced £140 which was paid into his bank in sovereigns so that he did not know his benefactors. Huxley organised a pension from the Civil List and a pension was also obtained from the Royal Society. So he was able to live in some comfort in Ventnor until his death ten years later.

Jones is an unusual instance of a physiologist who was alive and still active in 1876 but was not asked to join the Physiological Society (p. 262) and did not seek to join. Perhaps it was as well so; the new Society's chief activity was to dine together and the presence of Wharton Jones would not have been conducive to a happy atmosphere.

Westminster Hospital Medical School

Henry Power (1829–1911)

Henry Power's father was an army officer, from a long line of army men; his mother was from Whitby in Yorkshire. By the chance of his father's occupation, Henry was born in Nantes and was soon taken to Barbados, where he survived the hurricane of 1831: the barracks were destroyed and five men killed, but the child remained unharmed in his cradle. As a consequence of the life of an army officer, Henry attended a series of nine schools and his entry into medical school was with the purpose of becoming an army surgeon.

He was apprenticed to Mr Wheeler and lived with his master in Gracechurch St, near St Bartholomew's Hospital. Mr Wheeler interested him in Botany and he won prizes in that subject. He enrolled as a student at St Bartholomew's Hospital Medical School and, in an autobiography, Power testified to the quality of Paget's lectures in Physiology (p. 51). As a student he also became a close friend of Savory (p. 53). Power qualified Licentiate of the Society of Apothecaries in 1854, Member and Fellow of the Royal College of Surgeons also in 1854 and MB, London in 1855.

In the later stages of his student life, Power earned a living as Paget had done, by coaching students and by writing for various medical journals. He was a medical correspondent of the *Lancet* throughout his life. In 1851 he was appointed Demonstrator in Anatomy at the Westminster Hospital School, then lectured on comparative anatomy and physiology, until in 1856 he was appointed Assistant Surgeon and Ophthalmologist at the Westminster. In 1867 he became Ophthalmic Surgeon at St George's and in 1870 he returned to Barts as its first Ophthalmic Surgeon.

His main career was in ophthalmology; he was President of the Ophthalmological Society, 1890–3 and in 1880 and 1895 was Vice-President and President of the Section of Ophthalmology at the annual meetings of the British Medical Association. Also at the meetings of 1869, 1878 and 1879 he had been Secretary, Vice-President and President of the Section of Physiology.

It was Power's writing which brought him some recognition amongst physiologists. In 1869 he was the chosen translator of Stricker's *Histology*, in which Klein was also involved (p. 156), and this important book appeared in 1870 and 1872. Between 1864 and 1876 Power edited the seventh, eighth and ninth editions of

Carpenter's *Principles of human physiology* (p. 29); in Power's editions it was virtually a new book. It was succeeded in 1884 by Power's own student text, *Elements of human physiology*. He also revised the *Lexicon of medical terms*, an enormous work which appeared over the twenty years 1879–99.

Presumably because of his books, Power became examiner in the University of London with Huxley; in Oxford with Rolleston; in Cambridge with Foster and Humphry and in Durham with Philipson. From 1875 to 1886 he was often an examiner at the Royal College of Surgeons. He was thus well known to the founders of the Physiological Society and he was one of those invited to join at the first meeting.

In 1881 he became Professor of Physiology at the Royal Veterinary College where he became a much loved teacher, honoured by the gift of an easy chair when in 1904 he retired, succeeded by Brodie.

He retired to Whitby, the native town of his mother and where he had frequently visited his daughter. There he was himself nearly drowned in the tragic deaths of his daughter and granddaughter in 1898 and this was said to have been the event which soon caused his own retirement. He died in Whitby in 1911 of heart failure, which is said to have been initiated by the strain on his right heart of climbing the steps to the pairsh church on the Sunday two days before his death.

Henry Power's son, D'Arcy (p. 220) joined the Physiological Society in 1880, forming the first example where father and son were both members.

Middlesex Hospital Medical School

Campbell De Morgan, FRS (1811–76)

De Morgan became a student of University College, London in about 1830, when Bell was still professor there. When De Morgan qualified he became House Surgeon at the newly formed Middlesex Hospital of which Bell was a founder. In 1842 he was appointed assistant surgeon and in 1847 surgeon. Later, in 1865 he became Lecturer in Surgery and was particularly interested in cancer, writing a book *The local origin of cancer*, published in 1872. After his death a memorial to him took the form of additional beds for the treatment of cancer in a ward carrying his name.

At earlier stages of his career, he had been lecturer in Forensic Medicine from 1841–45 and in 1845 was appointed Professor of

Anatomy at the Middlesex, lecturing at first in descriptive anatomy. However for eighteen years, 1847–65, he gave three lectures per week on physiology and general anatomy; in 1865 he was followed by Burden Sanderson (p. 143) who for about six years was Assistant Physician at the Middlesex. In 1853, De Morgan published a paper on the Structure and Development of Bone and was elected FRS in 1865. Following Burdon Sanderson, for a short period the lectures in Physiology were given by Ferrier (p. 191).

Guy's Hospital

Guy's Hospital (Cameron, 1954) departed from the usual arrangement in that lecturers in General Anatomy and Physiology were usually physicians rather than surgeons. Before 1834 lectures specifically on physiology had been given by Haighton and then Blundell (p. 9). When, in a row with Harrison, the treasurer, Blundell gave up these lectures, the course in physiology was omitted and Guy's, like the other hospitals, had lectures in anatomy and physiology given by a surgeon, Bransby Cooper, nephew, assistant and successor to Sir Ashley Cooper. The lectures in physiology began again in 1846, partly, it has been said, because Harrison, the rather autocratic treasurer, wished to create a position for Gull, his protegé. For ten years lectures specifically in physiology were thus given by a physician, Gull, and he was succeeded by Pavy, also a physician. Pavy continued to give these lectures until 1877 and also established an experimental laboratory in Guy's; his life is therefore better considered in Part III (p. 209).

Sir William Withey Gull, FRS (1816–90)

Gull was born in Colchester, the son of a barge owner and wharfinger. His parents moved to Thorpe-le-Soken, a small harbour on an Essex estate owned by Guy's Hospital. He was educated privately as a pupil of the rector of the nearby village of Beaumont; the rector was a nephew of Harrison, the great treasurer of Guy's. For a short time Gull was a pupil teacher in the school of Mr Abbot at Lewes; Mr Abbot was a friend and correspondent of Michael Faraday and at this school Gull was trained in science, particularly botany. When Gull was eighteen he met Harrison, who was so impressed by the boy that he persuaded him to enter medical education at Guy's, entering him as a student without fees and providing accommodation. Gull did well, graduating MB London in 1841 and MD in 1846, and became Fellow of the Royal College of Physicans in 1848.

Gull had thus fulfilled Harrison's hopes and at Guy's was made

medical tutor and Lecturer on Natural Philosophy (1843−7) and then
Lecturer in Physiology and Comparative Anatomy, 1846−56. He was
also lecturer in Physiology at the Royal Institution, 1847−9. Gull
made no original contributions to physiology, although he was elected
FRS in 1869 in recognition of his contributions to pathology and
medicine.

Appointed Assistant Physician in 1851 and full Physician in 1858,
he was at first associated with Addison and was his immediate
successor as Lecturer in Medicine. Gull wrote many clinical papers
along the lines initiated by Bright and Addison. Of some physiological
interest was a paper in 1874 on a *Cretinoid state supervening in adult
life in women* in which he described what is now known as myxoedema.
As a physician he was noted for his dislike of the overuse of drugs
of dubious efficacy. He was a worthy successor to Bright and Addison
and become another great figure in the history of Guy's; particularly
he raised the status of students in medical wards, creating the name
of 'clerk', similar to that of 'dresser' in surgery.

He had a large private practice, which quickly increased after 1871
when he attended the Prince of Wales during an attack of typhoid
fever. For this Gull received a baronetcy and was appointed Physician
to the Queen and Prince of Wales. He was much honoured in official
posts of the Royal College of Physicians.

The London and Queen Square

John Hughlings Jackson, FRS (1835−1911)

Hughlings Jackson was a Yorkshireman, born at Green Hammerton
near Knaresborough and educated at local schools. His medical
education began early with apprenticeship to a practitioner in York
and he also was a student of the York Medical and Surgical School,
a small but good school at that time. He had already acquired much
of the teaching required for qualification when in 1855 he became
for a short time a student at St Bartholomew's Hospital and under
the influence of James Paget (p. 49). Jackson took the diplomas of
MRCS and LSA in 1856.

He returned to York and was appointed house surgeon to the dispensary, where his chief was Thomas Laycock, afterwards Professor
of Medicine in Edinburgh. After two years, in 1859, Jackson returned
to London and might well have abandoned medicine for literary
work but for the influence of Laycock and particularly Jonathan

Hutchinson, his senior by seven years and a Yorkshireman of very similar origins. Hutchinson became a life-long friend and helped Jackson to obtain appointment in 1863 as assistant Physician and Lecturer in Physiology at the London Hospital. In the meantime in 1860 Jackson had taken the degree of MD, St Andrews, and in 1861 had been admitted a Member of the Royal College of Physicians. At the London he lectured in physiology, pathology and the use of the microscope until 1870; in 1874 he was appointed full Physician, until his retirement from this post in 1894, with the title of Consulting Physician.

In parallel with his positions at the London, and of great importance to the science of Neurology, Jackson was in 1862 appointed assistant Physician to the National Hospital for the Paralysed and Epileptics, Queen Square, becoming Physician in 1867 until he retired in 1906. The hospital in Queen Square began in 1859; until 1864 Brown-Séquard had been one of its first physicians and Jackson's contact with him confirmed and directed a growing interest in the function and diseases of the nervous system. From 1870 onwards Hughlings Jackson was an acknowledge leader in the thinking of the group of great neurologists who collected at Queen Square.

The name of Hughlings Jackson is known to physiologists by his description of 'Jacksonian epilepsy', which was published in 1873. Jackson's method was accurately to describe clinical cases and in this way was able to recognise motor and sensory areas of the cerebral cortex, an area involved in speech, at a time when the idea of cerebral localisation of function was not yet established. Jackson's work preceded that of Ferrier (p. 190) and others who used experimental methods to establish local functions of the cortex. In his later life, Jackson also influenced the work of the next generation of neurophysiologists, Head, Horsley, Schäfer, Sherrington and Mott.

Jackson began publishing his descriptions of cases in 1864 and by 1902 had published over 200 papers mostly in the *Lancet, British medical journal* and other clinical journals, as listed by Broadbent (*Brain*, 1903, *26*, pp. 305–66). Jackson wrote no book reviewing his ideas. His findings and the deductions from them were sometimes grouped in papers under the title 'Remarks on ...' subjects such as chorea, epilepsy, etc. and he gave several formal lectures such as the Goulstonian Lectures (1869), the Hunterian Oration (1872), the Lumleian Lectures (1890). He was particularly interested in the ophthalmoscope and in 1885 gave the Bowman Lecture to the

Ophthalmological Society, which in 1897 founded the Jacksonian Lectures in his honour and he gave the first lecture in the series. In his early career Jackson was much influenced by the idea of evolution as expressed by Darwin and Huxley and, in a more general sense, by the theories of the philosopher Herbert Spencer (p. 114). This gave a philosophical bent to his writings which perhaps foreshadows the philosophical outlook of Sherrington.

Jackson's work was soon recognised by clinicians and scientists. He was elected Fellow of the Royal College of Physicians in 1869 and was immediately chosen to give the Goulstonian Lectures in that year. Later he was active and important in the affairs of the College. Jackson was elected FRS in 1878 and in later life was much recognised by honorary degrees of Universities and Colleges. Leeds conferred the degree of DSc in 1904 and this recognition in his own county gave Jackson great pleasure in the last years of his life.

Jackson cannot strictly be called a physiologist since he did no experimental work, although he did use the results of others in his arguments. He was elected a member of the Physiological Society in 1885 and remained a member until his death, although he took no part in the activities of the Society. Indeed he rarely, if ever, attended a meeting but this probably reflects his general attitude to any kind of official function. A sort of impatience could overtake him on any occasion. It is recorded that he would leave a train short of his destination and astonish the porter by asking urgently for a cab to carry him to the destination of the train from which he had just alighted. Attending a play he might sit through the first half and then leave, returning next day to see the second half, at the cost of a second ticket. Hutchinson, his life-long friend, tells many such stories in an obituary article (*BMJ*, 1911, ii, pp. 1551–4).

Dublin

Robert McDonnell, FRS (1828–89)

Robert McDonnell's grandfather and father were doctors in Dublin and Robert was born there. He entered Trinity College, Dublin in 1844 and was also apprenticed to Carmichael at the Richmond Hospital. He graduated in Arts in 1849 and Bachelor of Medicine in 1851 and he also passed Licentiate of the Royal College of Surgeons of Ireland; he was elected Fellow of the College in 1853. He travelled to study in Edinburgh, Paris and Vienna and in 1855 volunteered as a civilian surgeon in the Crimean War, being attached to British Hospitals at

Smyrna and Sebastopol. When he returned to Dublin he took MD in the University of Dublin in 1857.

In Dublin he became assistant surgeon in the Richmond Hospital and also lectured in anatomy and physiology at the Richmond (Carmichael) School of Medicine. At this time he did some research work, being particularly interested in the glycogenic function of the liver, a subject then being brought forward by the work of Claude Bernard. He was elected FRS in 1865. From 1857–67, he was medical superintendent of the Mountjoy Government Prison.

In 1866 McDonnell moved to Dr Steevens' Hospital where he was appointed assistant surgeon and Professor of Descriptive Anatomy. He was an extremely deft and skillful surgeon until ill-health forced him to resign in 1888. He had resigned his professorship in 1880. In this later phase of his life, McDonnell was a leading surgeon in Dublin and generally well known. In 1877 he was President of the Irish Royal College of Surgeons and in 1885–8 President of the Royal Academy of Medicine.

Well known in London, he was a foundation member of the Physiological Society in 1876 but published no papers in the *Journal of Physiology*. By then he had ceased experimental work; most of his extensive published work was in *Transactions and proceedings of the Royal Society*.

* * *

Augustus Volney Waller, FRS (1816–70)

Confusion can arise when A. V. Waller is not distinguished from his son, Augustus Desiré Waller, whose career as a physiologist is related on p. 215.

Augustus Volney Waller was born near Faversham in Kent but the family moved to Nice, where he grew up and was educated until his father died when he was fourteen. Augustus then went to live with his relations, the Lambe family, firstly in Tewkesbury and then with William Lambe, a physician in London. William Lambe was an eccentric who believed in two causes of disease states, either an animal diet or the impure water supplied in the metropolis. He was a vegetarian and A. V. Waller was thus brought up on a vegetarian diet until he returned to the continent for medical studies, obtaining MD (Paris) in 1840.

He then returned to England, qualified Licentiate of the Society of Apothecaries and began to practise in London at Kensington.

He began research work and published papers in the *Philosophical transactions of the Royal Society* and was elected FRS in 1851. However he returned to the continent to work with Flourens in Paris and Budge in Bonn; in Paris he also became known to Magendie and Claude Bernard. In 1858, for about a year, he was Professor of Physiology at Queen's College, Birmingham, but returned to the continent because of ill health. He worked in Paris, Bruges and finally in Geneva where he went into practice but died of *angina pectoris* in 1870. During this rather nomadic career his research work was recognised in Paris twice by the award of prizes of 2000 francs by the Academy of Science on the report of Claude Bernard and in London by a Royal Society Medal in 1860; he was also a Croonian Lecturer.

The main topics of A. V. Waller's contributions to physiology were set out by his son in the dedication of his textbook of 1891 *Introduction to human physiology* by A. D. Waller:

> To the memory
> of
> my father
> Augustus Waller, MD, FRS
> 1816–1870.
> > Emigration of Leucocytes, 1846
> > Degeneration and regeneration
> > of nerve, 1850
> > Cilio-spinal region, 1851
> > Vaso-constrictor action of sympathetic, 1853

Although his son always insisted on his father's discoveries in the other fields listed above, modern physiologists know of A. V. Waller only by virtue of 'Wallerian Degeneration'. Two points about this discovery are mentioned elsewhere in this book. Firstly (p. 55) it seems that Wharton Jones may have indicated the problem to Waller; they were both in London at the time. Secondly an account of the work was published in the *Transactions of the Royal Society*, as well as in Paris. Sharpey as referee recommended its publication but indicated that Waller had not mentioned some relevant existing observations (p. 82). It would be interesting to know more of what Waller did in London about 1850 and who his contacts were, but the obituary notices deal only briefly with this period.

Sir Benjamin Ward Richardson, FRS (1828–96)

Benjamin Ward Richardson was born at Somerby, Leicestershire, the son of Benjamin Richardson and his wife, Mary Ward; both families had long lived in the district. His education was begun by his mother, who destined him for the medical profession, even to her dying words. After her early death his education was continued in the school of a local clergyman, Mr Nutt, until he was apprenticed to a Somerby surgeon, Henry Hudson. Both of these teachers imbued him with a love of natural science and Hudson particularly involved him in experiments with electrical and mechanical devices.

In 1847 Richardson went as a medical student to Anderson's College, Glasgow, where he also attended the lectures of Robert Knox who, driven out of Edinburgh by the Burke and Hare scandal, had then set up a private school of anatomy in Glasgow. However, Richardson fell a victim to a form of recurrent fever and withdrew from Glasgow for reasons of health; he served as an unqualified assistant to practitioners at Saffron Walden, Narborough, and then Barnes, where he had access to London libraries, scientific appliances and teachers in science. His master at Barnes was Dr Robert Willis, editor of the Medical Gazette and at this time Richardson became acquainted with Jerrold, Lemon, Cruickshank and Thackeray, then engaged in establishing *Punch*. For the rest of his life Richardson was a figure in literary circles.

However he was not yet medically qualified, so in 1850 he returned to Glasgow to pass Licentiate of the Faculty of Physicians and Surgeons and moved on to St Andrews to become in 1854 MA and MD of that University. Richardson was much attached to St Andrews; for the rest of his life he sought to establish its status and it was chiefly due to him that the University of St Andrews obtained representation on the General Medical Council and he formed the St Andrews Medical Graduates Association, of which he was President for thirty-five years until his death.

From 1854, Richardson was established in medical practice in London at 25 Manchester Square and was also privately a very active experimentalist in matters relevant to medicine. Although he held only minor posts as a physician in hospitals, he achieved in the next fifteen years very wide recognition as a practitioner and also as an experimenter. The most striking tribute to his fame was in 1868, when over 600 scientific, medical and literary men contributed a purse of over 1000 guineas and a microscope by Ross, which was presented to him at a meeting of 400 of the subscribers, presided over by James Paget (*Lancet*, 1868, i, p. 670). This was a remarkable tribute to a man who held no appointments at the great hospitals and taught very little. He was elected Fellow of the Royal College of Physicians in 1865 and was invited to deliver some of its formal lectures, such as the Croonian Lecture of 1873. He was elected FRS in 1867.

His teaching was in the private medical school in Grosvenor Place, near to St George's Hospital. In 1854 he began lecturing in medical jurisprudence and after 1857 he changed to lecturing in anatomy and physiology and in hygiene; his predecessor as Lecturer in Physiology had been the father of Ray Lankester, Edwin Lankester, who like Richardson had been a pupil assistant of Dr Thomas Brown of Saffron Walden. For a time Richardson was Dean of the Grosvenor Place Medical School but it closed in 1865 by merger with nearby St George's. Thereafter Richardson did no formal teaching.

Richardson's experimental work was on topics related to medicine rather than pure physiology. His most widely acclaimed work about 1870 was on anaesthesia. He introduced some fourteen substances as possible alternatives to chloroform or ether; he invented a double-valve inhaler for the more accurate administration of chloroform but appreciated that deaths resulted from the use of volatile anaesthetics and sought an alternative. Thus he developed the idea of anaesthesia by local cold and invented a device for spraying ether on to the surface, to produce local anaesthesia under which surgeon friends carried out major operations such as amputations and Caesarian Section. Also he developed procedures and apparatus for painless killing of domestic and farm animals. In 1868 the RSPCA presented him with a medal in recognition of this work and also the introduction of low temperature anaesthesia for operations on domestic animals. In his speech in presenting the purse, Paget (see above) said that Richardson would be remembered as a chief alleviator of pain.

Richardson also worked on the coagulation of blood, which he thought involved emission of ammonia; the formation of fibrinous concretions in the heart; palpitation and intermittent action of the heart; tuberculosis and scarlatina. His wide-ranging work also included studies of the actions of amyl nitrite and various alcohols and he introduced a number of possibly useful medicinal agents.

Alongside his scientific work, Richardson had much influence in matters of public health, particularly as a sanitary reformer in matters of domestic detail, and was an advocate of bicycling. In 1875 in a celebrated address at Brighton entitled 'Hygeia', Richardson described what a city should be if sanitary science were properly applied. As a private venture, he started in 1862 the *Journal of public health and sanitary review* to support his aphorism 'National Health is National Wealth'. After four years this was succeeded by *Social science review*, also short-lived, in which Richardson included some of his literary efforts such as poems and plays. At the end of his life he was producing *Asclepiad*, a quarterly publication containing original research, observations and criticism in science, art and literature of medicine in its widest sense. Richardson was a most prolific writer in fields of medicine and literature.

Scientific description of the effects of alcohol convinced Richardson that alcohol was dangerous and not to be taken except under medical prescription. He became a rigid abstainer and was an active advocate of temperance. His Cantor Lectures of 1875 became a standard authority in the cause of temperance.

His knighthood in 1893 recognised his standing as a man of importance in humanitarian affairs rather than as a pure scientist. Richardson's last book was of an autobiographical nature, giving his thoughts on the many facets of his life. *Vita Medica; chapters of medical life and work'* was published by his son after his death; he had completed it only hours before he was struck by a cerebral vascular accident from which he died.

In 1876 when the Physiological Society was founded Richardson accepted the invitation to be a founder member, but apparently took no part in the affairs of the Society. He seems never to have attended a meeting – perhaps his temperance ill matched the function of the Society as a dining club. His death was reported to the Society in 1896.

Always interested in matters of medical history, during his life Richardson wrote a series of appreciative biographies which were published posthumously under the title *Disciples of Aesculapius* (1900). As preface, the book contains a Life of the Author written by his daughter. In his and his daughter's eyes, Richardson's most important work was his contribution to public health, fulfilling his mother's dying command that 'he must not only make sick people well, but must keep the people that were well from becoming sick'.

References

Broadbent, W. (1903). 'Hughlings Jackson as pioneer in nervous physiology and pathology', *Brain*, **26**, 305–66.

Cameron, H. C. (1954). *Mr Guy's Hospital 1726–1948*. London: Longmans Green.

Godlee, R.J. (1921). 'British Masters of Ophthalmology Series. 12: Thomas Wharton Jones' (P), *British Journal of Ophthalmology*, **5**, 97–117, 145–56.

Moore, N. (1918). *The History of St. Bartholomew's Hospital*, Vol. II. London: Pearson.

Newman, C. (1957). *The Evolution of Medical Education in the Nineteenth Century*. Oxford University Press.

Paget, S. (1901). *Memoirs and Letters of Sir James Paget, edited by one of his sons*. London: Longmans Green.

Sharpey-Schafer, E. (1927). *History of the Physiological Society during its first Fifty Years, 1876–1926*. Issued by the Society and published as a supplement to *Journal of Physiology*, December. Cambridge University Press.

Biographies

Baker:	*British Medical Journal* (1896), ii, 1169–70 (P).
	Moore (1918), 688–91.
Gull:	*Lancet* (1890), i, 324–6.
	Cameron (1954), 233–4 (P).
Jackson:	*British Medical Journal* (1911), ii, 950–4 (P).
	Broadbent (1903).
Jones:	*Lancet* (1891), ii, 1256–8.
	British Medical Journal (1891), ii, 1175–7.
	Godlee (1921) (P).
Kirkes:	*British Medical Journal* (1864), ii, 714–5.
	Moore (1918), 567–8.
McDonnell:	*Lancet* (1889), i, 965.
	Dictionary of National Biography, **35**, 59.
De Morgan:	*British Medical Journal* (1876), i, 523–4.
Paget:	*Lancet* (1900), i, 52–6 (P).
	British Medical Journal (1900), i, 49–54 (P).
	Paget, S. (1901) (P).
Power:	*British Medical Journal* (1911), i, 233–4 (P).
	Lancet (1911), i, 274–7 (P).
Richardson:	*British Medical Journal* (1896), ii, 1612.
	Lancet (1896), ii, 1575–6 (P).
Savory:	*Lancet* (1895), i, 648–9 (P).
	British Medical Journal (1895), i, 564–5.
	Moore (1918), 680–2.
Waller, A.V.:	*Lancet* (1870), ii, 489.
	Nature (1870), **2**, 436.
	Dictionary of National Biography, **59**, 122–3.

6

The only full-time physiologist – Sharpey at University College, London

University College, London (1828–36)

A University of London, an idea often proposed, came to fruition in 1828. It was to provide University education in a wide range of subjects, with the special provision that there would be no religious tests. At that time students of Oxford and Cambridge were required to accept the Church of England. The University of London was not a free university; the students were to pay fees which would provide income for the professors and the University, but it was hoped that the costs to the students would be less than at Oxford or Cambridge. Medical courses were included in the original foundation, because there were many medical students in London who could be attracted to the new University and so contribute an immediate income; fifty-four students attended the first course in anatomy.

In 1826 the site of the present University College and University College Hospital was acquired and building began. In 1827 a group of professors was nominated, including three professors of Anatomy, Physiology and Surgery, one of whom was Charles Bell (p. 3). On 1 October 1828 Bell gave the first lecture, describing himself as Professor of Surgery and Physiology and this was also the inaugural lecture in the whole University. According to the *Lancet* (1828–9, *15*, 8) the lecture was given to a distinguished audience of about 600 in a room to the north of the central dome of the main building of University College; the dome was then still being built.

The University of London in its original form did not last long. Almost immediately, in 1829, King's College, London was set up with the support of the Archbishop of Canterbury, as an alternative to the 'atheist' establishment in Gower Street. Thereafter there were two colleges, University College and King's College, formalised in 1836 when charters were granted to the two colleges as teaching institutions. A separate charter set up the University of London with the role of holding examinations and granting degrees; its degree of Bachelor of Medicine gave the holder the right to practise as a qualified doctor. At this time the established hospitals, St Thomas's

Guy's, St Bartholomew's and the London, had no connection with the University of London.

At first, when University College had no charter to grant a licence to practise medicine, the lecture courses were advertised as recognised to provide certificates required for examination at the Royal College of Surgeons or Apothecaries' Hall. There were lectures on medicine and surgery but no hospital in which clinical instruction could be given. In 1834 the North London Hospital, soon to be called University College Hospital, was opened and clinical instruction could then be given. Immediately the number of medical students increased from 248 to 353.

The *Lancet* of 26 September 1829 lists the courses available for the session 1829–30. Those relevant to physiology were:

Anatomy: Prof. Pattison. 1.30–3.00 daily, except Saturday.
   Fee £7. (The half-hour from 1.30–2.00 will be occupied by examination on the previous lecture.)
Physiology: Prof. Bell. 5.00–6.00, Tuesday and Thursday.
   Fee £2.
Anatomical Demonstrations: J.R. Bennett, daily, 11.00–12.00.
   Fee £5. (Anatomy of the human body will be covered twice in the session.)
Comparative Anatomy: Prof. Grant. 3.00–4.00 daily, except Saturday.
   Fee £2. (For one term only.)
Surgery and Clinical Surgery: Prof. Bell. 5.00–6.30 Monday, Wednesday, Friday.
   Fee £5.
Museum of Anatomy is open every day.

Examinations and certificates
Every professor devotes a certain portion of the hours of instruction in each week to the examination of his pupils. No student is exempted – all requiring certificates must submit to these examinations.
There will be three public examinations of each class in the course of the session; the first immediately before Christmas; the second immediately before Easter; the third at the beginning of July, or middle of May for medical students.
At the third examination prizes and honours will be awarded.

The note 'examinations and certificates' does not appear in the notices of courses in other schools and is an indication of a growing belief that teaching institutions should provide supervision of the students enrolled in their courses.

From the beginning there was strife amongst the professors of the new university. Three professors, Bell, Pattison, and J.P. Meckel of Halle were nominated to share the teaching in anatomy and physiology, morbid and comparative anatomy and surgery. Meckel never accepted appointment and Bell and Pattison, disliking each other, shared out the first lectures as shown in the list on the previous page. In 1831 a separate Professor of Surgery was appointed. *Samuel Cooper* (1780–1848) had been a student at St Bartholomew's Hospital, becoming MRCS in 1803. He served as an army surgeon at Waterloo and became one of the first surgeons of University College Hospital. He was highly regarded but resigned in an atmosphere of distrust, fully ventilated in the *Lancet* (1848, i). *Bell* resigned in 1830 (p. 5). *G. S. Pattison* (1791–1851) never got on with Bell and was accused by the students of being incompetent. There was much to support this accusation, such as his appearing for lectures in hunting dress and not attending the dissecting room. After much official wrangling and a series of near riots by the students, Pattison was dismissed in 1831; he was subsequently Professor at Jeffearson College in Philadelphia and New York. *J. R. Bennett* was regarded by the students as a very good teacher and indeed was well qualified by a period in Paris as a teacher in Anatomy. For a short period he succeeded as Professor but died in 1831.

For a short period lectures in Anatomy were given by *Thomas Southwood Smith* (1788–1861) who was a remarkable man. At the foundation of the University in 1828 he was considered for but not elected as Professor of Moral Philosophy. He was subsequently an important figure in the introduction of public health measures in England. Born in Somerset, Southwood Smith became an evangelical minister in the West Country, encouraged by William Blake, before proceeding to Edinburgh in 1812 to obtain medical qualification; he became MD in 1816. In the meantime he continued his evangelical ministry and can be regarded as a founder of a Scottish Unitarian Church. Southwood Smith set up in practice in London, taking the qualification of Licentiate of the College of Physicians in 1820 and began to interest himself in matters of public health. He wrote a *Treatise on fever* (1830), *The philosophy of health* (1836) and in 1829 *Treatise on animal physiology*. With the sudden resignation of Bell, Southwood Smith filled the gap by giving lectures in anatomy for a period in 1831. Southwood Smith has a further connection with University College because he was the friend who was instructed by the will of Jeremy Bentham to dissect Bentham's body, as the subject of a public lecture. This was done on 9 June 1832 at the Webb Street School, Southwood Smith speaking firmly through a thunderstorm, 'but with a face as white as that of the dead philosopher before him'. Southwood Smith prepared the skeleton, mounted on it a wax likeness of Bentham's head and clothed the skeleton in a suit of Bentham's clothes; he kept this effigy in his rooms for a number of years before presenting it in its cabinet to University College (Marmoy, 1958).

In 1831, University College appointed Jones Quain as a new Professor with the title Professor of General Anatomy. *Jones Quain* (1796–1865) was a graduate of Trinity College, Dublin, in both arts and medicine, and had visited continental schools before settling in London as lecturer in the Aldersgate School, then of high standing under the ownership of W. Lawrence. His brother, Richard Quain, was already at University College as a demonstrator but he and Jones Quain were not friendly and Richard established himself as independent of the direction of his brother and had reached independent professorial status before Jones Quain retired in 1835. Jones Quain, who was later a member of the Senate of University College, London, apparently retired because he could not be bothered by the squabbling amongst the professors. He was interested in literature and occupied no further academic or medical posts, living chiefly in Paris.

After 1836, a stable arrangement was reached in the teaching of anatomy and physiology. When Jones Quain resigned, a committee of professors was asked to recommend to the Council what steps should be taken to fill the vacancy. The committee recommended that two professors of equal standing should be appointed – Dr Richard Quain to teach descriptive anatomy and Dr William Sharpey to teach anatomy and physiology. A similar separation was made at King's College (p. 35), and a little later at St Bartholomew's (p. 51). This partial separation of anatomy and physiology was not yet the recognition of physiology as a separate subject. Rather, the purely descriptive anatomy of the dissecting room and applied to surgery was separated from general anatomy, function and morbid anatomy; physiology as we know it had not yet begun. However, the appointment of Sharpey as Professor of Anatomy and Physiology at University College London is usually regarded as the beginning of English physiology. He held his chair with this title from 1836 until his retirement in 1874 and during those years and under his influence physiology was established as a separate subject. When he died in 1880 the essentials of modern physiology had been reached.

The appointment of Sharpey (1836)

Tradition of the Physiological Society, perhaps originating from Sharpey's devoted pupil, Schäfer (who later adopted the name of Sharpey-Schafer,

(p. 147) suggests that the appointment of Sharpey was by unanimous decision. Taylor (1971) has studied the records of University College, which reveal that the decision was reached by narrow majorities and after much manoeuvring, with R. Quain promoting Sharpey's interest and recommending to Sharpey that he come to London. A deciding factor seems to have been Sharpey's arrival in London with nine excellent references from notables in Edinburgh. These references still exist, all in Sharpey's handwriting.

The report of the committee was not welcomed by the *Lancet*. In an editorial of 13 August 1836, Wakley did not regard Sharpey as the best candidate:

The professors have advised the conjoint appointment of Mr. R. Quain, and a Doctor Sharpey of Edinburgh to the chair of Anatomy ... Here we have Dr Grant who is, beyond all dispute, one of the most highly-gifted physiologists in Europe, and whose reputation has extended everywhere, made, *by his own colleagues*, to give place to a Doctor Sharpey, who has not the felicity of being known out of Edinburgh!

Grant was already Professor of Comparative Anatomy, a post which he continued to hold. Nor did the *Lancet* like the idea of two equal professors: 'No sooner will those gentlemen be placed together in the same chair, than they will each endeavour, under the common impulses of their nature, to sidle one another off their seats.' Finally the *Lancet* did not like the way the appointments were made. It wanted some sort of popular election and regarded the committee of professors as a kind of self-electing establishment and Wakley regarded all establishments as inherently corrupt: '... in the instance before us we have hypocrisy, treachery, envy and fraud, superadded to the one ancient evil, *love of pelf*' (*Lancet*, 1836, *30*, 789). Despite the *Lancet*'s forebodings, Sharpey, Quain and Grant worked amicably together. The *Lancet* of 30 September 1837 listed lectures at University College in 1837−8 as;

Anatomy and Physiology	Prof. Sharpey	Daily at 12	£6
Descriptive Anatomy and Dissections	Prof. R. Quain	Daily at 10	£6
Comparative Anatomy	Prof. Grant	Daily at 3 except Thursdays and Saturdays	£4

and these were continued until Sharpey's retirement in 1874, except that after Quain's removal to surgery in 1850, the lectures on 'Anatomy, descriptive and surgical' were given by Mr Ellis. During this long period of professorial stability, University College became recognised as the best school of anatomy in London and the only school of physiology.

Anatomists at University College

Richard Quain, FRS (1800–87)

The Quains were an Irish family from County Cork. Jones and
Richard were brothers and a younger Richard Quain (p. 88) was a
cousin. Richard Quain, the anatomist, served a medical apprentice-
ship in Ireland before attending lectures at the schools in Windmill
Street and Aldersgate where his brother Jones was lecturer in
Anatomy. Richard passed MRCS in 1828. Earlier in 1825 he had been
in Paris, firstly as pupil, then assistant in Anatomy to J. R. Bennett,
who when the University of London opened in 1828 was appointed
Demonstrator in Anatomy with Richard Quain as his assistant. When
Bennett was promoted and then died, Quain became Senior
Demonstrator. When Jones Quain came in as Professor of General
Anatomy, Richard insisted that he be independent of his brother and
remain in charge of the dissecting room, his position having been
recognised in 1832 by official appointment as Professor of Descriptive
Anatomy. When Jones Quain resigned in 1835, Richard retained his
position of Professor of Descriptive Anatomy and was able to
influence the appointment of Sharpey as his co-professor.

In the meantime the North London Hospital had been opened and
developed into University College Hospital. Liston was brought from
Edinburgh to be Surgeon and in 1834 Quain was appointed Assistant
Surgeon. He became Full Surgeon in 1850 and resigned from his chair
of Anatomy to devote his time to surgery and his post of Professor
of Clinical Surgery (1848–66). He was elected Fellow of the Royal
College of Surgeons at the first such election in 1843 and thereafter
held many official positions in the college. He was nominated as its
representative on the General Medical Council from 1870–6. He
became Surgeon Extraordinary to the Queen in 1878 and with Sir
James Paget (p. 49) attended the final illness of G. H. Lewes (p. 118).

Richard Quain was a typical surgeon-anatomist of his time, who
enters the story of physiology by his association with Sharpey. Jones
Quain began *Quain's Elements of Anatomy*; for the fifth edition in
1843 the authors were R. Quain and Sharpey with Sharpey writing
the section on general anatomy (see p. 28). Quain's most important
book as an anatomist surgeon was *The anatomy of the arteries of the
human body, with its application to pathology and operative surgery*
(1844). He was elected FRS in 1844.

Richard Quain wrote several articles on general and medical

education and certainly made his contribution to the development of University College. An obituary notice in the *Lancet* (1887, ii, 687) says that he was a popular lecturer, clear and dogmatic and implies that in committee he was difficult, obstinately holding opinions formed in advance. In his will a fortune of £75,000 was left to University College, London, for furthering education in language and science. By this legacy Quain Studentships were founded and a Professorship in English Language and Literature was endowed.

George Viner Ellis (1812–1900)

Born near Gloucester, Ellis was apprenticed to a doctor in Gloucester and then entered University College, London, qualifying MRCS in 1835; he was elected FRCS at the first elections in 1843. Soon after qualifying he became Demonstrator in Anatomy at University College under Richard Quain and also spent vacations working in anatomy in Paris or Berlin. He never attempted to become a surgeon but remained in anatomy at University College, succeeding Quain as Professor of Descriptive Anatomy in 1850 and retiring in 1877. He held no official posts outside University College. He retired to Gloucester where he taught older boys in a night school and applied himself to gardening, particularly growing apples. He died in 1900, having been blind for the last two or three years of his life.

Ellis was perhaps the first pure anatomist. His lectures made no attempt to make the dry subject more interesting but were well accepted by students because they stated clearly what had to be learnt for examinations. He was a firm disciplinarian in his own classes and was once called in to quell misbehaviour in Grant's class. He hated smoking. Smoking in the dissecting room was firmly forbidden until a group of students petitioned the Council for leave to purify the atmosphere of the practical anatomy rooms with tobacco; the corpses at that time were not preserved. Ellis immediately threatened to resign and the sanction was rescinded. He was not overclean in his habits and it was not always appreciated, when he showed his approval for an industrious piece of dissecting by patting the student on the shoulder with hands 'too visibly subdued to that they worked in'. In a college epitaph he was described as:

Beloved by few and feared by many,
Discoverer of the corrugator ani.

Early in his career he published *Demonstrations of anatomy* and together with Sharpey he was editor of *Quain's Elements of Anatomy* after 1856. With Ford, an artist, Ellis published in 1867 *Illustrations of dissections* in two volumes, one volume consisting of plates drawn by Ford from dissections by Ellis.

Robert Edmund Grant, FRS (1793–1874)

Robert Edmund was the seventh son in a family of fourteen children of Alexander Grant of Edinburgh, writer to the signet, and was a pupil of the High School before entering the University of Edinburgh as a medical student in 1808. He qualified LRCSE and took the degree of MD in 1814. He then travelled widely in Europe before returning to Edinburgh in 1822. The *Lancet* (1850, ii, 686) gives details of his remarkable travels by which he became highly qualified in languages, zoology and anatomy. He did some private medical practice in Edinburgh but mainly devoted himself to zoology and anatomy, giving lectures on the comparative anatomy of invertebrates.

The years 1822–7 in Edinburgh were his period of activity as an original worker; particularly he published a classical paper on sponges, recognising them as living and describing the circulation of fluids through their structure. He was elected FRSE.

Well recognised in Edinburgh, Grant was appointed Professor of Zoology and Comparative Anatomy at the foundation of the new University of London and he retained this chair in University College, London until his death in 1874. In London he produced some small notes on the structure of various animals but his output of original work was very small. He was elected FRS in 1836, and in 1837–40 was Fullerian Professor at the Royal Institution.

At University College he devoted himself to his course of lectures, giving four or five lectures per week for forty-six years, never missing a lecture. Grant's lectures of 1833–4 were published in the *Lancet* (Vol. 25) and they were very good lectures at that time. However Schäfer, who attended the lectures in 1869, stated that Grant did not change them and was delivering the same lectures thirty or forty years later. Lectures on comparative anatomy were not necessary for examinations and audiences fell, the class at times becoming as few as two students (and these were not regular in their attendance).

Grant's is a pathetic story of unfulfilled promise. Well educated, a master of European languages and widely travelled, he was a failure in his University life. He was retiring and sensitive yet capable of arousing friendship. Originally he was very poor (professors were then paid by the fees from the students who enrolled for their lectures) and friends had to establish an annuity of £50 per annum for him. Later he inherited money and was dying intestate until his friend Sharpey persuaded him to leave his books and money to the library of University College; the Grant Library was arranged by Sharpey himself.

Obituary notices in *Nature* and the *Dictionary of National Biography* tell of his failure. His only consistent supporter was the *Lancet*. At the time of the appointment of Sharpey, the *Lancet* vehemently advocated the appointment of Grant as Professor of Physiology (p. 73) and indeed his achievements at that time would have justified the appointment. Thereafter the *Lancet* always regarded him as a man who had been badly treated and in 1850 published a laudatory 'Biographical Sketch of Robert Edmund Grant, MD, FRS' (*Lancet*, 1850, ii, 686–95). This account of his early life is in contrast to the assessments of others.

Grant was known to both Darwin and Huxley and was in some degree a forerunner in the ideas published in 1859 in *Origin of species*. In the *Life of Charles Darwin* by his son, Francis, Darwin wrote an autobiographical chapter of his early life, in which he says of his acquaintances in Edinburgh in 1825:

Lastly, Dr Grant was my senior by several years. He published some first rate zoological papers, but after coming to London as Professor in University College, he did nothing more in Science, a fact which has always been inexplicable to me. I knew him well; he was dry and formal in manner, with much enthusiasm beneath this outer crust. He one day, when we were walking together, burst forth in high admiration of Lamarck and his views on evolution ... Dr Grant and Coldstream attached much to marine Zoology and I often accompanied the former to collect animals in the tidal pools, which I dissected as well as I could.

Similarly, Huxley wrote:

Within the ranks of biologists in 1851–8, I met with nobody, except Dr. Grant of University College who had a word to say for evolution – and his advocacy was not calculated to advance the cause. Outside these ranks, the only person known to me whose knowledge and capacity compelled respect, and who was at the same time a thorough-going evolutionist, was Mr. Herbert Spencer (p. 114), whose acquaintance I made, I think, in 1852 and then entered into the bonds of a friendship which has known no interruption.

William Sharpey, FRS (1802 – 80)

William Sharpey was born at Arbroath. His father, a ship owner, had gone there from Folkestone; his mother, Mary Balfour, was an Arbroath woman and William was the last of their five children, his father dying before he was born. Mary Balfour married a second time, to Dr Arrott of Arbroath, and had further children. William was thus brought up in the home of a Scottish doctor and went to school in Arbroath. In 1817 he went to the University of Edinburgh to attend classes in Greek and natural philosophy and in 1818 commenced studies in the University and extra-academic medical schools in Edinburgh, obtaining his diploma of the Edinburgh College of Surgeons in 1821. He further studied anatomy in Brookes' School in London and in 1821 – 2 he spent a year in the wards of Paris Hospitals with Dupuytren, before taking his Doctorate of the University of Edinburgh in 1823. Then for five years he was walking in Europe. At his death his biographers did not know much about how he travelled but he spent long periods at those medical schools which had well-known anatomists, such as Panizza in Pavia and Rudolphi in Berlin.

On returning to Edinburgh, he lectured in anatomy in the extra-academic school in association with his younger friend Allen Thomson (p. 103), who also studied anatomy by travelling on the continent. Somewhat unexpectedly, in this association, Sharpey gave the lectures on anatomy and Thomson those on physiology. Their classes rose from twenty-two to eighty-eight, so Sharpey was a successful lecturer but was not in fact outstandingly successful. Robert Knox in 1828 was lecturing to 500 students in Anatomy and his classes remained large even after 1828, when he had become notorious as the anatomist who had bought bodies from Burke and Hare. However Sharpey did obtain the reputation of being a good lecturer and also a good scientist, which was the basis of his appointment to University College in 1836. Equally his achievements had not been such as to make his appointment automatic. Although Sharpey had been elected FRSE in 1834, there was some justification for the *Lancet*'s assertion that he 'has not the felicity of being known out of Edinburgh'.

However he was appointed in 1836 and for thirty-eight years until his retirement in 1874, Sharpey was Professor of Anatomy and Physiology at University College, London; he kept this title although over the years he came to be recognised as a physiologist. Early in his career he had decided to devote his life to anatomy and physiology and so never took any clinical appointments but devoted his whole time to his academic work of lecturing and administration. The main work was the course of lectures, some 120 per year, as set out in the list on p. 73, and much of his influence on the development of physiology came through these lectures. From his position of a senior professor, Sharpey made great contributions to administration in University College and the University of London, to the General Medical Council and was also secretary of the Royal Society at a time of great changes. It is uncertain how much research work Sharpey did. Altogether Sharpey's professional life, discussed in more detail in following paragraphs, was very similar to that of a professor and head of department of today; it was unique in that he was the first full-time Professor of Physiology.

By the standards of any period Sharpey was a very good Professor. Among his colleagues he was accepted as completely honest in what he did and his opinions and decisions had to be respected. His contact with students was through lectures to large classes but he still managed to make contact with individual students. Sharpey-Schafer (1927) recalled 'He had the usual gift of being able to recognise each student of his class and to remember his name, although hundreds attended his lectures. Not only did he know them, but he took pains to learn all about them and could recall their circumstances and career even after many years'. Obituary notices of Sharpey all speak of his attractive personality.

Towards the end of his time at University College, in gratitude for his teaching and friendly help, former pupils founded the Sharpey Memorial Scholarship at University College while he was still professor, and also presented a portrait in oils and a bust. The scholar was to be assistant to the Professor and the first Sharpey Scholar was appointed in 1869 – it was E. A. Schäfer, who later changed his name to Sharpey-Schafer (p. 147). Sharpey himself contributed to the scholarship fund and in his will left a further £800. He also left the University his collection of books.

Sharpey remained in full vigour until in 1870 he developed cataracts. Despite operations he was considerably handicapped and after 1870 some of his lectures were given by Burdon Sanderson (see p. 141). Sharpey decided to retire in 1874 and was granted a government pension of £150 per annum. He still lived in London to be a senior consultant on such matters as the Antivivisection Act in 1876, and at the foundation of the Physiological Society was elected an Honorary Member.

Sharpey died in London in 1880 and was buried in the cemetery adjacent to the ruins of Arbroath Abbey. Londoners paid their last respects to a much loved and respected man on 15 April 1880 when his body was carried *en cortege* from University College to Euston Station followed by fifty private carriages and a vast number of people on foot.

The *British Medical Journal* (1880, i, 637) gave this account of the funeral at Arbroath on Saturday, 17 April.

The remains, which arrived on Friday, were conveyed to the house of Dr Sharpey's half-sister, Miss Arrott. Previously to leaving for the Abbey Burying-ground, a religious service was performed by the Reverend George Logan of Inverbrothock. The funeral was a private one; but, as a mark of respect for the deceased, most of the shops along the line of the route which the mourning-coaches took were closed, and the steeple bell tolled at intervals. On arriving at the Abbey gate, the coffin was removed from the hearse and borne to the grave ... A large crowd of spectators thronged the streets and collected in the church yard; among them were not a few aged persons whose memories carried them back to the early days of their distinguished townsman, whose past career had shed such lustre on his birthplace, and whose personal qualities endeared him to all who knew him.

The *British Medical Journal* names the pall bearers and others who were present; they include Dr James Arrott of Dundee (Sharpey's half-brother) and citizens of Arbroath, Dundee, Broughty Ferry and Edinburgh. Sharpey's London friends and colleagues were not present.

At Arbroath he was buried in a family grave. The inscription on the stone, standing in the burial ground adjacent to the Abbey ruins, begins:

Set up in 1873 by
WILLIAM SHARPEY, M.D.
in the place of a stone then fallen into decay
which bore the following inscription.

On its two sides the stone records that a stone was first set up by MARY BALFOUR in memory of her husband, HENRY SHARPEY, who died in 1801 and of her two sons, HENRY and DAVID, who died in 1805 and 1808, aged seven and eight respectively. On the reverse side is the name of MARY BALFOUR, widow of HENRY SHARPEY, who married Dr Wm ARROTT of Arbroath and died in 1836, aged 62. Also recorded are the deaths of their two daughters, ELIZABETH, wife of Wm Colvill, who died in London in 1855 and is buried in Highgate Cemetery, and ISABELLA, wife of Major George Goodall, who died at Heidelberg in 1861. Finally there is just room for:

WILLIAM SHARPEY, M.D., F.R.S.
Professor of Anatomy and Physiology,
University College, London
son of the above
HENRY SHARPEY and MARY BALFOUR
born 1st April, 1802, died 11th April, 1880
and was buried here.

Sharpey did not marry. His house in Torrington Square, London was kept by his niece, Miss Colvill, until she died in 1878. Of his father's family, Sharpey was survived by only one nephew, William Colvill, who soon died. Of the Arrott family, his step-father died in 1862 and Sharpey was survived by a half-sister in Arbroath and a half-brother, a doctor in Dundee. The name of Sharpey ended with William Sharpey, but is perpetuated in the memory of physiologists because Sharpey's pupil, friend and colleague, Schäfer, changed his name to Sharpey-Schafer (p. 147).

Sharpey's administrative work
As a senior professor at University College, Sharpey was necessarily involved in its administration. He was Dean of the Medical Faculty in 1840−2 and was re-elected for two futher terms in 1852−4 and 1862−4. The Dean was involved in many matters of decision in University College Hospital ranging from drains to the appointment of senior staff, who were also Professors in University College. Sometimes such appointments involved violent arguments with, as was usual at that time, letters of personal abuse in the *Lancet*. Occcasionally when Sharpey was forced to answer such letters, he did so effectively and with dignity.

From 1840−63 he was examiner in anatomy and physiology in the University of London and was later a member of the Senate. From 1860 to 1875 he was a member of the General Medical Council, for a time its treasurer, and so was influential in the changes in medical education which took place at that time. From 1870 to 1875 he was a member of the Commission on Education and Science, and he was also a trustee of the Hunterian Museum of the College of Surgeons.

He was elected Fellow of the Royal Society in 1839 and in 1844–5 was a member of the Council. In 1853 he became one of its two secretaries and, throughout his period of office until 1872, he was the active secretary. During those twenty years the Royal Society began to operate as it does today in respect of the publication of its transactions and proceedings, the way in which it elects new fellows and the cataloguing of scientific papers; in 1857 the Society moved from Somerset House to Burlington House. Sharpey is honoured as an important figure at this stage of the history of the Royal Society. Indeed for a time he was something of an autocrat in the affairs of the Society and so drew some personal attacks. In a matter concerning an election to the Council, Huxley disagreed with Sharpey, but after discussion deferred to Sharpey's judgement, confident of his sense of fairness and justice.

Sharpey wrote many reports on candidates for membership and, in effect, referee's reports on papers submitted for publication. These reports were of high quality, properly critical but always fair. Amongst the papers he refereed was A. V. Waller's paper on the degeneration of nerves (p. 64); Sharpey recommended its acceptance but pointed out that the author was apparently unaware of some existing work on this subject.

Sharpey resigned official positions at the same time as he resigned his chair, about 1874, largely because of failing sight and hearing. However his experience and great authority was much utilised, for example in the enquiry into Antivivisection and in representations on the Bill which was passed in 1876. He obviously supported the foundation of the Physiological Society, was present at the founding meetings and was elected an Honorary Member. He attended meetings until 1878.

Research work by Sharpey

Sharpey published very little original research. While in Edinburgh he published his observations on cilia in a number of invertebrate animals, *On a peculiar motion excited in fluids by the surfaces of certain animals* (*Edinburgh medical surgical Journal*, 1830, *34*, 113–22) and he followed this in 1835 by an article which is essentially a translation of work of Purkinje and Valentin but with his own confirmation of their work. From his review of the literature, Sharpey indicates that his paper in 1830 was not in fact the first description of ciliary movement. In Edinburgh, a paper on blood vessels in the porpoise and infusoria was also largely translation of the work of European authors with some additions. He was elected fellow of the Royal Society of Edinburgh in 1834.

At University College it seems that Sharpey used the microscope and carried out some experiments to confirm published findings which he intended to use in lectures or articles. Thus Sharpey wrote the chapter on Bone, Osseous Tissue in the fifth edition of Quain's *Elements of Anatomy* (p. 28); this article for many years was the accepted description of the histological structure of bone. It is like a review of today but it is impossible to know how much of original observations by Sharpey is included. Most of what he wrote for Quain's *Anatomy* is of the same type and he similarly wrote sections for Todd's *Cyclopaedia* (p. 36). Taylor (1971) examines the existing records of Sharpey's lectures with the conclusion that Sharpey did some experimental work to examine what he wished to include in each lecture. He never published any of this work as formal papers.

One is forced to conclude that Sharpey's contribution to Physiology was not as an original discoverer; he was a superb and critical reviewer and this was important at the time when original literature was not freely available.

Sharpey's Lectures

For the first twenty years of his professorship, Sharpey's only formal contact with his students was through his lectures given to an audience of about 200. In later life his students were able to say that they had been attracted by his lectures; Lister, who attended Sharpey's lectures in 1848, said in 1900 'as a student at University College I was greatly attracted by Dr Sharpey's lectures which inspired me with a love of Physiology which has never left me'. Students of that time were perhaps more attuned to lectures than the students of today. It is not easy to understand how any student would not get bored by the same lecturer five days a week for a whole session.

Schäfer said, 'In person, Sharpey was large and stoutly built reminding one of an enlarged edition of Mr Pickwick and he had a habit when lecturing of keeping his left arm rigidly to his side or somewhat behind him which was also suggestive of that creation'. In 1852, W. S. Jevons (later Professor of Logic, Owens College, Manchester), wrote home, 'Sharpey is a very nice old fellow, and one of the best physiologists alive. He attended Mr Graham's class this year with all the other students, and since Easter has been working all day in the laboratory, with Dr. Williamson telling him how to do the things. You must not complain of me making messes and blow-ups in the cellar if an old chap of sixty begins to learn to do it'. Graham was Professor of Chemistry, and Williamson of Analytical and Practical Chemistry, at University College and Sharpey attended their full courses in 1852. Herein, probably, lay Sharpey's success with students; he gave the impression of studying and learning with them, and sharing his conclusions with them. He was not a superb lecturer in the manner of Paget or Bell, but he was a friendly and effective teacher.

For thirty-eight years, from 1836 to 1874, Sharpey lectured under the same title, anatomy and physiology. Over almost the same period, from 1828 to 1874, Grant lectured under the title of comparative anatomy and physiology. Whereas Grant's lectures never changed in forty years, it seems that Sharpey's lectures were continually changed to include new work as it came out, mostly from Germany, Holland or France. A feature of Sharpey's lectures which interested better students was that he dealt with new work, describing the findings and giving his critical assessment of its meaning and value. In an obituary notice of R. Quain in the *Lancet* (1887, ii, 687), an ex-student of Quain and Sharpey contrasts their approach. Sharpey, the critical physiologist, would introduce his doubts about a point – 'You see, gentlemen, in a manner, so to say, the phenomena ...' whereas Quain, the surgical anatomist, was more dogmatic: 'The fact is, gentlemen, ...'.

Taylor (1971) refers to four sets of students' notes from Sharpey's lectures which still exist in manuscript – by Potter (1836–7), Ballard (1840–1), Lister (1849–50) and Thane (1867–8). From Potter's notes, Taylor (1971) gives two examples of Sharpey's inclusion of recent work. Schwann (1833, 1836) used the term 'pepsin' for a ferment in gastric juice: in 1836–7, Sharpey had apparently carried out experiments to confirm and extend Schwann's findings, demonstrated them to his class and discussed their significance. But in Ballard's notes of 1840–1 pepsin is only briefly mentioned. Similarly in Potter's notes of 1836–7, the observations of Beaumont (1833) on the gastric mucosa of Alexis St Martin are considered, although at that time Beaumont's work was not well known or widely accepted in Europe.

The *Lancet* (Volume 39) published the first five lectures of Sharpey's course in 1840–1. These lectures do not read particularly well; they are rather formal, oratorical, with a verbosity which was then usual. However they are probably not typical in that they were written out. Biographers all state that Sharpey lectured without notes, apparently ex tempore, although in fact the lectures had been carefully prepared, and much work had gone into preparing illustrative material or demonstrations.

In the published introductory lectures of 1840–1, Sharpey outlines the course. He stated that descriptive anatomy would be dealt with in the parallel course by Quain and his own lectures would deal with general anatomy and function. General anatomy was the precursor of histology. Sharpey started from the idea enunciated in about 1800 by Bichat; the individual organs were made up of different admixtures of a few common 'textures', what we call gland cells, muscle cells, connective tissue, blood vessels, nervous tissue. In 1840 microscopes available were insufficient to establish this by direct observation, and so the speculations of Bichat were still cited. Later lectures dealt with function.

According to Taylor (1971) new developments in physiology were immediately incorporated into Sharpey's lectures, as in the example of pepsin just cited. By 1850 in Lister's notes, descriptive histology had replaced general anatomy as the compound microscope came into use. Thane's lecture notes in 1867–8 contain full discussion of the experimental physiology from recent experimental work mainly on the continent from such physiologists as Claude Bernard, du Bois Reymond and Ludwig. The up-to-date element in Sharpey's lectures had much to do with his power to attract good students and interest them in physiology. Sharpey's lectures were the basis of a book by Marshall (p. 88) *Outline of physiology, human and comparative*, published in 1867. Marshall had been a student of Sharpey in 1840, so this book may not truly represent Sharpey's lectures of the 1860s.

Practical classes in histology and experimental physiology

In 1862 Sharpey gave a progress report on the development of physiology in an 'Address in Physiology' to the 30th Annual General Meeting of the British Medical Association; he was also reporting on his own thirty years as a physiologist. He said (*Lancet*, 1862, ii, 182–8) that physiology had in recent years received general recognition amongst educated people and as a subject in schools. Sharpey then recounted some recent advances in physiological knowledge, ascribing them to three factors. Firstly, there were now instruments letting us see things: he mentions the ophthalmoscope and laryngoscope but most importantly the increasing availability of compound microscopes; by 1862 there was accurate histology of most tissues. Secondly, there were now numerical measurements of physiological function and he cites the measurement of arterial blood pressure, circulation time, blood volume, and conduction velocity in nerve. Thirdly, animal experimentation was being properly used, especially on the Continent, although there was as yet very little experimental work in Britain. Sharpey did not mention the increasing knowledge of organic chemistry.

These points in the advance of physiology had already been recognised in Sharpey's teaching by his introduction at University College of practical classes for students, although these were not compulsory. The announcement in the *Lancet* of 20 September 1856 of the courses at University College in the following session included classes in histology daily in the summer term at 3 pm, fee £2. Prior to 1856 Sharpey had been demonstrating microscopic preparations to his students, using a circular table fitted with a railway line on which a single microscope could be passed to each student in turn. Similar devices were used by Acland at Oxford and Quekett at the Royal College of Surgeons. Also, since about 1853, there were classes in histology conducted in Edinburgh by Hughes Bennett (p. 96). Presumably there were, at University College in 1856, sufficient microscopes and a room was made for the small class in histology by screening off part of the dissecting room. In 1856 G. Harley (p. 151) was appointed Lecturer in Practical Physiology and Histology; according to Bellot (1929) this lecturership was instituted 'with the view of supplying the Medical Students with instruction in the use of the microscope in examining the textures and fluids of the body'. In 1859 the title of the class given by Harley was expanded to 'Histology and Experimental Physiology', given in the winter term on Mondays and Wesnesdays at 4 pm and Saturdays at 10 am, fee £3. A room on the first floor at the north end of the college, originally the medical library, was fitted up for experimental physiology and histology. Harley's classes were written by one of his students into a book on histology and a second edition was produced by Harley himself.

However it seems that the course did not develop greatly in Harley's time. When he resigned in 1867, Michael Foster (see p. 167) was appointed firstly as Instructor and then, in 1869, as Professor in

Practical Physiology and Histology; more students took the course and a new physiological laboratory was established on the ground floor to the north side of the entrance block of the college. Years later, Foster recalled the situation when he first went to University College (*Johns Hopkins Hospital Bulletin*, 1911, *22*, 329):

But what could be done then was very, very little. I had a small room. I had a few microscopes. But I began to carry out the instruction in a more systematic manner than had ever been done before. For instance, I made the men prepare the tissues for themselves. That was a new thing in histology. And I also made them do for themselves simple experiments on muscles and nerves and other tissues on live animals. That, I may say, was the beginning of the teaching of practical physiology in England.

Sharpey's successors and pupils

Foster, Sanderson, Schäfer, Martin

When, in 1870, Foster went to Cambridge, Burdon Sanderson was appointed Professor of Practical Physiology and Histology, with Sharpey, now in doubtful health, still Professor of Anatomy and Physiology and Ellis as Professor of Descriptive Anatomy. This represents the final achievement of Sharpey's career; from his tentative introduction thirteen years earlier of practical classes for students, he had established in physiology two younger and vigorous men who in the next ten years established Physiology as an experimental subject to be taught in a laboratory as well as by lectures. Foster and Burdon Sanderson were the active immediate founders of two great schools of experimental physiology in Cambridge and in University College, London.

As assistants they had two younger men brought up under Sharpey's influence. Schäfer (p. 147) stayed at University College and succeeded to the chair when Burdon Sanderson went to Oxford. Martin (p. 171) accompanied Foster to Cambridge and then went to Johns Hopkins University as its first Professor of Biology.

Sharpey has been called the 'Father of English Physiology', a title given him by his immediate successors. To later generations, who knew him not, the justification for this title is not immediately obvious. He made no original discoveries and he formed no school of experimental physiology like those of Claude Bernard in Paris or Ludwig in Leipzig. What he did do, by his teaching, character and example, was to introduce into physiology remarkable pupils and successors. Physiology was indeed fortunate that in 1870, when its growth became possible, it had Foster, Sanderson and Schäfer and our tribute can correctly go to their common teacher, Sharpey.

Foster and Sanderson could in the 1870s quickly develop physiological laboratories because so many of those then in influential positions had either been direct students of Sharpey or had become his friends — 'he knew everyone and was known by everyone'. In the following paragraphs are short biographies of such men, influential in 1870, who were early pupils of Sharpey. Sharpey's pupils were, of course numerous and his influence correspondingly great; he taught classes of 200 for thirty-eight years.

Sir Richard Quain, FRS (1816—98)

This Richard Quain, like his two older cousins, Jones (p. 72) and Richard (p. 74) came from County Cork. He was apprenticed to an apothecary in Limerick; in 1837 he proceeded to University College Medical School where his cousin Richard was then Professor of Anatomy. He graduated MB London in 1840 with a gold medal in physiology. He served for about five years as house-physician and became physician of Brompton Hospital where he worked for his whole life.

Quain became an important figure in the Royal College of Physicians, member in 1846, fellow in 1851 and later member of the Council and Censor, gave several orations and in 1889 was vice-president. His biggest contribution was as a member of the General Medical Council; appointed as a Crown nominee in 1864, he was President from 1891 until his death and was particularly concerned in the production of editions of the *Pharmacopoeia*.

He was appointed to a Royal Commission on cattle plague on which he was associated with physiologists, Burdon Sanderson (p. 141), Marcet (p. 157) and Beale (p. 42). He was elected FRS in 1871 and in 1861 became a member of the Senate of the University of London. He was a very successful physician, appointed physician extraordinary to the Queen and knighted in 1891.

His most important writing was the *Dictionary of medicine*, written with collaborators over the years 1875—82.

John Marshall, FRS (1818—91)

Marshall was born at Ely and apprenticed to a surgeon at Wisbech. His apprenticeship was apparently not very successful and in his later interest in medical education, he had no wish to retain the apprentice system. He entered University College as a student in 1839 and passed Member of the Royal College of Surgeons in 1844. At University College he was Curator of the Anatomical Museum and demonstrated in Anatomy under Sharpey and Quain, until he was appointed

Assistant Surgeon at University College Hospital and later full surgeon; he was Professor of Surgery from 1866–85.

The obituary notice in the *Proceedings of the Royal Society* says that as a student he attracted the attention of Quain and Sharpey and that he remained thereafter a close friend of Sharpey. Certainly he retained interest in physiology. He was four years Fullerian Professor at the Royal Institution and also Lecturer in Anatomy as applied to Art at the Government School of Art at South Kensington. Here he found it was possible to lecture to mixed classes of men and women without giving offence.

In 1850 and 1864 he published papers in the *Philosophical Transactions of the Royal Society* on the embryology of veins in man and mammals and on the brains of idiots, and was elected FRS in 1857. Marshall published books on anatomy and physiology – *The human body: its structure and functions* (1860) and *Anatomy for artists* (1878). *Outline of physiology: human and comparative* (1867) was based on Sharpey's lectures. In 1876 he was a founder member of the Physiological Society.

His eminence as a surgeon gave him important influence in the College of Surgeons, elected to the Council in 1873 and President in 1883. Being interested in medical education, he much influenced the setting up of the conjoint examination in 1884. He was the representative of the College of Surgeons on the General Medical Council after 1883 and, succeeding Quain, was president of the council from 1887 until his death.

Marshall was much interested in art. He was intimate with Maddox Brown, to whom he sat as the Jester in the painting, 'Chaucer'. He was also friend and doctor to D. G. Rossetti.

Marshall was buried in the public cemetery at Ely. His widow presented a memorial window to Ely Cathedral. Executed by Henry Holiday, the window is based on the theme of Christ healing the sick and is fringed with edelweiss, Marshall's favourite flower (*Lancet*, 1893, ii, 1646).

John Eric Erichsen, FRS (1818–96)

Erichsen's father was Danish and he was born in Copenhagen, but his mother was from Somerset and he lived all his life in London. As a medical student at University College he came under the strong personal influence of Sharpey and passed Member of the College of Surgeons in 1839. For a brief period he taught physiology at the

Westminster Medical School. His career was in surgery, however, and after 1848 he became in succession Assistant Surgeon and Surgeon at University College Hospital, Professor of Surgery from 1850–66 and Professor of Clinical Surgery from 1866–71. His main published work was an important book, *The science and art of surgery* (1853).

At the College of Surgeons he became Fellow in 1845, a member of the Council in 1869, Examiner in 1875 and President in 1880. He was appointed surgeon to the Queen. He was President of University College for the nine years before his death.

Erichsen's importance to physiologists lies in his being a member of the Royal Commission of Enquiry into Vivisection (1875) and later home office inspector under the act of 1876. Possibly these official positions prevented his joining the Physiological Society. He was elected FRS in 1876.

Erichsen, on a holiday at Pontresina in Switzerland in 1865, met Mrs Gaskell, the novelist, and her daughter 'Meta'. Despite the tonsillectomy performed by Bowman in 1861 (p. 39), 'Meta' was in poor health and, in a letter to a friend, Mrs Gaskell said that Erichsen took a kindly interest and prescribed for her – open air, early hours, plenty of meat, bitter beer, warm sea douches.

Joseph, Baron Lister of Lyme Regis, FRS (1827–1912)
Joseph Lister was born at Upton, Essex into a Quaker family, which came originally from Bingley in Yorkshire. His father, Joseph Jackson Lister, FRS (1786–1869), was a merchant in London who applied himself to the study of optics and to whom we owe the achromatic lens and so to a large degree the compound microscope. He encouraged Smith to engineer suitable instruments to use his lens (p. 45) and one of these good microscopes was provided for his son. In other ways also Joseph was brought up in a scientific and literary background, became interested in biology and announced his intention to be a surgeon.

In 1844, Joseph entered University College, London, graduated BA in 1847 and after a delay due to an attack of smallpox, became a medical student in 1848, qualifying MB London and taking Fellowship of the Royal College of Surgeons in 1852. After two years in University College Hospital as a house surgeon, he was advised by Sharpey to go to Edinburgh for wider surgical experience with Syme, to whom he became trusted assistant and friend. Well recognised in Edinburgh, in 1860 he was appointed Professor of Surgery in Glasgow where he stayed for nine years. Then Syme was forced to retire from ill-health and Lister was appointed in 1869 as Professor of Surgery in Edinburgh. Next, in 1876 he was approached by some of the staff of King's College, London, including Bowman, and after some negotiation, was appointed to a specially created chair: so in 1877 Lister went to King's as Professor of Clinical Surgery, from which he retired because of age limit in 1893.

From about 1865 in Glasgow, he was developing the use of carbolic acid in treating wounds and in surgery. Combined with cleanliness this revolutionised surgery. At the same time Lister improved other techniques of surgery, introducing the use of absorbable catgut for ligatures. The results of Lister's methods were so spectacularly improvements that the old ways were displaced in all countries within about twenty years, despite some conservative opposition to innovation. At the International Congress in London in 1881 an opponent was W. S. Savory (p. 53), then a senior surgeon at St Bartholomew's Hospital.

For fifteen years after his retirement from King's, Lister was a great world figure, a senior statesman in the world of medicine and science and the recipient of a vast number of honours in many countries. He was knighted in 1884 and raised to the peerage in 1897, taking the title of Baron Lister of Lyme Regis. Elected Fellow of the Royal Society in 1860, he served as Foreign Secretary, before becoming President of the Royal Society, 1895–1900. In 1902, at the coronation of Edward VII, Lord Lister was appointed a first member of the newly-created Order of Merit.

Before he became involved in his new techniques in surgery, Lister carried out original research of high quality on physiological topics. The work of 'Lister as a Physiologist' was set out by Sharpey-Schafer (1927). As a medical student in 1849, Lister's great ability had been recognised by Sharpey, who thereafter became his friend and counsellor. Lister attended Sharpey's lectures twice and his manuscript notes of the lectures of 1849, modified in 1850, are the most complete record still extant of Sharpey's lectures (see p. 84). Next Lister was assistant to Wharton Jones (p. 54) in University College Hospital and many years later in his full fame, Lord Lister paid tribute to the influence of these two physiologists.

Lister's first research work, while still a student, was made possible by the microscope provided by his father; as with Bowman (p. 37) the direction of his work was determined by the possession of a microscope. In 1853 he published a careful description of the smooth muscle of the iris (modifying Bowman's earlier paper) and also of the smooth muscle of the skin. These papers were published while he was house surgeon at University College Hospital and were certainly discussed with Sharpey. In Edinburgh Lister continued histological and experimental work, publishing papers on the minute structure of the smooth muscle, the cutaneous pigmentary system of the frog, the regulation of blood vessels in the frog's skin by nerves, coagulation of the blood, the function of visceral nerves with special reference to the so-called 'inhibitory system'. These papers were all published in the period 1854–9 and included experiments on animals, sometimes without anaesthetic; in his later fame, Lister was an uncompromising supporter of the need for experiments on animals. Amongst this work, his famous paper on the early stages of inflammation was largely microscopic observations, in continuation of the work of Wharton Jones (p. 54). This paper is regarded as the beginning of Lister's ideas on the nature of inflammation which, combined with his appreciation of the work of Pasteur, led to antiseptic surgery and treatment of wounds.

Physiologists recognised Lister as one of themselves by electing him in 1892 as Honorary Member of the Physiological Society and in 1912 the minutes recorded the death of 'a very distinguished Honorary Member, Lord Lister'. He does not appear to have attended meetings of the Society.

James Blake (1815–93)

Blake was born at Gosport near Portsmouth and in 1886, at the age of 71, was elected a member of the Physiological Society. Little is known of his career except that he was a student under Sharpey at University College, London, about 1839. About 1845 he went to America, firstly to St Louis, then Sacramento and finally settled in San Francisco. He became Professor of Obstetrics at Toland Medical College, established in 1864.

Published papers indicate that he attempted some research. The first group of papers are in the *Edinburgh medical and surgical journal* and the *Lancet* in 1839–41. They are reports of experimental work that he did while a student at University College, London; he acknowledged that an idea was derived from Sharpey's lectures and stated that some of the experiments were sanctified by the presence of Sharpey. The experiments consisted of studying the effects on the circulation of substances injected intravenously in dogs and other animals using an early form of mercury manometer. The articles are dated from University College or a private address in London, except for one from Paris with acknowledgement to Magendie, Serres and Flourens. In this period he signed himself as MRCS (England).

In the *Journal of Physiology*, 1884—6, are four papers dated from San Francisco restating some of his earlier work and commenting on current work, including some work of Ringer. In 1886—7 Blake gave three communications to meetings of the Physiological Society, joined the Society and presided at a meeting in 1887 at King's College. In 1888, in a letter to the *Lancet* on the climate of California, he wrote from Paris and signed himself MD (London), FRCS.

Finally he returned to San Francisco and died at Middleton, handicapped by an ununited fracture of the thigh.

References

Bellot, H. H. (1929). *University College, London 1826—1926.* University of London Press.

Marmoy, C. F. A. (1958). 'The "Auto-Icon" of Jeremy Bentham at University College, London', *Medical History*, **2**, 77—86.

Sharpey-Schafer, E. (1927). 'Lister as a Physiologist', Ch. III in *Joseph, Baron Lister; Centenary Volume 1827—1927* (edited by A. L. Turner) (P). Edinburgh: Oliver and Boyd.

Sharpey-Schafer, E. (1927). *History of the Physiological Society during its First Fifty Years 1876—1926.* Published as a supplement to the *Journal of Physiology*, December.

Taylor, D. W. (1971). 'The life and teaching of William Sharpey (1802—1880): "father of modern physiology" in Britain', *Medical History*, **15**, 126—53, 241—59.

Biographies

Bennett:	*Lancet*, 1831, **20**, 149, 403—4.
Blake:	*Dictionary of American Medical Biography*, edited by H. A. Kelly and W. L. Burrage (1979). Longwood Press: Tortola.
Cooper:	*Lancet* (1848), ii, 646.
Ellis:	*British Medical Journal* (1900), i, 1132.
Erichsen:	*British Medical Journal* (1896), ii, 885—7 (P). *Proc. Roy. Soc. London* (1897), **61**, i—iii (by Schäfer).
Grant:	*Lancet* (1850), ii, 686—95 (P). *Nature* (1874), **10**, 355—6.

Lister: *Proc. Roy. Soc. London B* (1913), **86**, i–xxi (P).
                              Lancet (1912), i, 465–72 (P).
                              Sharpey-Schafer (1927) (P).
Marshall: *Proc. Roy. Soc. London* (1890–1), **49**, iv–vii.
                              Lancet (1891), i, 117–19.
Pattison: *Dictionary of National Biography*, **44**, 58.
Quain, J.:   *Dictionary of National Biography*, **47**, 89–90.
Quain, R. (anatomist):   *Lancet* (1887), ii, 687–8.
Quain, R. (physician):   *Lancet* (1898), i, 816–20 (P).
                              Proc. Roy. Soc. London (1898), **63**, vi–ix.
Southwood Smith: *Dictionary of National Biography*, **53**, 135–7.
                              Marmoy (1958) (P).
Sharpey: *Lancet* (1880), i, 662–4.
                              Proc. Roy. Soc. London (1881), **31**, x–xix.
                              Taylor (1971) (P).

7

Physiologists in Scotland

Institutes of Medicine

Throughout the nineteenth century the usual relationship between physiology in Scotland and England was that young men, after medical education in Edinburgh, moved to London. Bell, Sharpey, Burdon Sanderson, Harley and Marcet took to London the experience of having been taught by the great anatomists of Edinburgh and had also some introduction to the science of medicine from the lectures of the Professor of the Institutes of Medicine. 'Institutes' included some clinical teaching and the professor was a physician dealing with patients, but with an outlook towards pathology and physiology. When a chair of the Institutes of Medicine was founded in Glasgow in 1839, it was also called the 'Theory of Physic'; often the Professor of the Institutes of Medicine was loosely called Professor of Physiology. By this system, physiology in Scotland was not subservient to anatomy, as it was in England before 1870.

In Edinburgh from 1848–74, the Professor of the Institutes of Medicine was Hughes Bennett; he had been preceded for a short period by Allen Thomson. Bennett is important as a founder of Physiology because he introduced in Edinburgh the teaching of physiology by practical classes. Two other professors in Edinburgh, Christison and Maclagan, founded experimental laboratories in their own subjects.

In Glasgow from 1850 Buchanan was Professor of the Institutes of Medicine but did nothing to advance physiology. After 1848 Allen Thomson was Professor of Anatomy and was the driving force in the formation of an effective medical school. Although in the period 1850–70, medical education in Glasgow was newly organised (Coutts, 1909), there was no real teaching in physiology until Buchanan was succeeded by M'Kendrick in 1875.

Similarly, organisation of medical education in Aberdeen began with the fusion of the colleges in 1860 (Simpson, 1963). The lecturer in the Institutes of Medicine over this time was Ogilvie who was a general biologist and physiology was not systematically taught until he was succeeded by Stirling in 1877.

Both M'Kendrick and Stirling had been assistants in Edinburgh, so that the advance in Physiology in Scotland can all be attributed indirectly to Hughes Bennett.

Physiology in Edinburgh

John Hughes Bennett (1812 – 75)

Bennett, born in London, attended school in Exeter, but the main influence in his education was his mother, who was of independent thought and brilliant intellectual powers. She cultivated his literary and artistic tastes, particularly causing him to read aloud speeches from Shakespeare with proper emphasis and gesture. He thereby acquired histrionic power which in later years gave elegance and finish to his lectures and speeches. He and his mother also lived in Europe, especially in France.

Bennett commenced the study of medicine as an articled pupil of Mr Sedgewick, surgeon at Maidstone. Zeal outran discretion when he conducted a *post mortem* without the consent of his master and his apprenticeship was abruptly terminated. In 1833 he enrolled as a student in Edinburgh although he knew no one in that city. Bennett was an assiduous student, particularly impressed by Robert Knox, the anatomist, and John Fletcher, physiologist, and he made friends with fellow students many of whom became eminent in medicine and science. He qualified MD in 1837 with the highest honours and a gold medal for a thesis on *The physiology and pathology of the Brain; being an attempt to ascertain what portions of that organ are more immediately connected with motion, sensation and intelligence.*

He then spent two years in Paris, devoting his attention particularly to the use of the microscope and to clinical work in the hospitals. There he learnt the new Paris approach to medical examination and diagnosis (see p.22), which he subsequently taught in Edinburgh.

In 1841 he returned to Edinburgh, with no appointment, and began a course of private lectures in histology. The advertisement of this course is of interest in relation to the development of teaching using the microscope:

Dr. Bennett, during the summer session, will give a public course of lectures on the Minute Structure of Organised Tissues, with reference to Anatomy, Physiology, Pathology and the Diagnosis of Disease. These lectures will be illustrated by numerous preparations, diagrams, and demonstrations under the microscope; the latter by means of twelve achromatic instruments of great power manufactured by Chevalier of Paris expressly for this course. The lectures will be on Monday, Wednesday and Friday at 11 a.m. throughout the session; fee £2.2s.

Dr. Bennett will also give private courses in the Practical Manipulation of the Microscope. Each class is limited to *six* and the time of the lecture regulated by the wishes of the majority. The lectures embrace the optical and mechanical arrangements of microscopes, illumination, mensuration, optical illusions, mode of displaying objects, and every information necessary for the medical enquirer, in his examination of the animal textures in a state of health and disease; fee £3.3s. 16 Pitt Street, May 2nd, 1842 (*BMJ*, 1875, ii, 473).

This was the first defined course of histology in Britain and was far more advanced than anything then available in London.

In 1842 Bennett became Fellow of the Royal Society of Edinburgh and of the College of Physicians in Edinburgh and physician to the Royal Dispensary. There he taught the methods he had learnt in Paris, particularly insisting on proper examination of the patient to lead to a diagnosis based on facts. From 1842 he was active in every aspect of medical life, giving lectures, conducting his classes, editing and writing for a private journal *Edinburgh monthly journal of medical sciences*. In 1848, when Allen Thomson went to Glasgow, Bennett was appointed Professor of the Institutes of Medicine in the University of Edinburgh.

Institutes of Medicine included physiology, but everything was to be taught in relation to pathology and therapeutics. Bennett interpreted his post to be that of a physician, but one who based his teaching and practice on physiology and pathology and in his later years viewed with dismay the tendency to divorce physiology from its practical relations to the wants of the medical profession. For twenty-five years he was a great teacher of precision in everything to do with medicine. He differed from many of the great lecturers of his time in that his lectures were fully written out before he gave them. This ensured precision, but his pupils recalled their joy, when he laid aside his manuscript and, *ex tempore*, attacked the opinion of an adversary.

In physiology he introduced practical classes, as he later described (*Lancet*, 1874, ii, 534):

Histology, systematic and practical, was first taught by me in 1841. Shortly after my appointment to the chair of Physiology in 1848, I saw the necessity of extending the practical department. But it was not until the year 1859 that

I with great difficutly succeeded in obtaining a grant of money for the purchase of instruments necessary for that purpose. This was followed by another grant, with which I was enabled to convert the premises previously occupied by the Theological Library, into a new theatre, a chemical and physiological laboratory, and an instrument room. These were completed in 1862. By the kindness of Prof. du Bois Reymond of Berlin, Prof. von Bezold, of Jena, visited Edinburgh and personally instructed me in the use of the many delicate instruments which had now arrived, and with which I performed numerous experiments in my systematic class. The next step was obtaining an annual salary for an assistant. This was granted in 1863, and Dr. Argyll Robertson was nominated to that office. This gentleman was the first to teach the three branches of the subject fully, under my direction – *viz*. Practical Histology, Practical Physiological Chemistry and Practical Experimental Physiology. During the winter session 1864–5 the class numbered 75 students.

Dr. Robertson informs me that immediately before his appointment (i.e. as assistant) Dr. Rutherford attended his class, and was instructed by him in the use of all the instruments, very much, as I believe, Dr. M'Kendrick in turn learnt from Rutherford.

Of these assistants Douglas Argyll Robertson (p. 99) became a famous ophthalmic surgeon, Rutherford (p. 198) was Professor at King's College London until in 1874 he succeeded Bennett and M'Kendrick (p. 202) became Professor of Physiology at Glasgow. M'Kendrick, however, stated that the instruments were imported by Goodsir, the Professor of Anatomy in Edinburgh, who forced Bennett to take them over (*BMJ*, 1909, i, 252).

Bennett was a prolific writer, producing some 105 publications, mostly on clinical topics or medical education. There was no original work in physiology but he published teaching books: *Outlines of physiology* (1858) and *Text-Book of physiology – general, special and practical* (1872).

About 1868 his health began to fail and he spent the winter session in Nice, his duties being carried on by Argyll Robertson. Again in 1872 and in 1873–4 he had to leave his duties to his assistant, now M'Kendrick, and finally he was forced to resign his professorship in 1874, to be succeeded by Rutherford. He died in Norwich in 1875.

At his retirement Bennett was honoured by the degree of LL D of the University of Edinburgh and a bust was presented to the University to be placed in the Library Hall. Twenty-five years later, money donated by his daughter allowed Schäfer to expand the facilities of the Physiology Department by new laboratories given the name of 'The Hughes Bennett Laboratories' (Turner, 1933).

Douglas Argyll Robertson (1837–1909)

Douglas Argyll Robertson (as he was always known) was born in Edinburgh; his full name was Douglas Moray Cooper Lamb Argyll Robertson. His father was a well-known surgeon in Edinburgh, a lecturer in the Extra-academical School, specialising in ophthalmic surgery. Douglas was educated at the Edinburgh Institution and in Germany and then in the Universities of Edinburgh, St Andrews and Berlin. He graduated MD St Andrews in 1857 and became Fellow of the Royal College of Surgeons, Edinburgh in 1862.

The next stage in his career is that which interests physiologists. In 1864 Argyll Robertson was appointed assistant to Bennett in the Institutes of Medicine at Edinburgh. His work was to conduct the newly-formed practical classes; for several sessions he taught chemical analysis, a short course in electrophysiology, the physiology of nerve and muscle, and histology. He instructed in these methods his two successors, Rutherford and M'Kendrick (*BMJ*, 1909, i, 252). Argyll Robertson did no research work in physiology.

Argyll Robertson, determining to follow his father in a career of ophthalmology, became in 1867 assistant in Ophthalmic Surgery at the Royal Infirmary and in 1870 full Ophthalmic Surgeon, until his retirement in 1897. He had studied with von Graefe in Berlin and was also influenced by Bowman. In his speciality, Argyll Robertson studied Calabar Bean as an agent in ophthalmology; its physostigmine-like action was then being described in the Department of Materia Medica by T. Frazer. Another contribution was his description of the pupil in *tabes dorsalis* responding to accommodation but not to light, thereafter called the Argyll Robertson pupil. He was lecturer in diseases of the eye in the University of Edinburgh, became like his father President of the Royal College of Surgeons of Edinburgh and was much honoured in the world of British ophthalmology, frequently president of societies and congresses. He was Surgeon-oculist to the Queen in Scotland.

Argyll Robertson was a very handsome and popular figure in the social world of Edinburgh. He was a prominent amateur golfer, winning the Gold Medal of the Royal and Ancient Club five times between 1865 and 1873. When he left Edinburgh, at a farewell speech in one of the clubs, Argyll Robertson dwelt on the value of golf to the busy professional man who knew how to make the wise and correct mixture of work and play. He was also keen and skilled in shooting, archery, fishing and curling. He left Edinburgh in 1904 to live in the warmer climate of Jersey.

While teaching in Edinburgh Argyll Robertson became a close friend of the Thakar of Gondal and was later entrusted with the care and education of the Thakar's eldest daughter. In 1909 Argyll Robertson and his wife travelled with the Princess to India, where Argyll Robertson was taken ill and died in Gondal. His body was cremated on the banks of the river Gondli, after the funeral service had been read by Rev. G. F. Stevenson; the Thakar himself kindled the funeral pyre and Hindus and Mussulmans closed their shops as a mark of respect (*Lancet*, 1909, i, 656).

Sir Robert Christison (1797–1882)

Robert Christison was the son of a Professor of Latin in the University of Edinburgh. He received a classical education at the High School until he was fourteen and then in the University of Edinburgh, before he turned to medicine and graduated MD Edinburgh in 1819. After serving as house surgeon and physician at the Royal Infirmary, he travelled to study medicine further at St Bartholomew's Hospital in London and then to Paris to be a pupil of Robiquet in chemistry and Orfila in toxicology.

Christison was still in Paris when the Chair of Medical Jurisprudence in Edinburgh became vacant; in his absence he was proposed by his friends and elected to the vacancy. Medical jurisprudence was then a rather neglected subject to which he brought some drive and turned it into a precise subject with his class increased from twelve to ninety. At this time he published a classical book *Treatise on poisons* and was a prominent witness in criminal cases, including the case of Burke and Hare.

In 1832 he transferred to the Chair of Materia Medica which he held for forty-five years. In addition to teaching this subject, he also taught clinical medicine but finding the income inadequate, he later undertook and built up a very large private practice and became regarded as a leading physician in Edinburgh. He was recognised by official positions, became Physician to the Queen of Scotland and was knighted. Above all he was devoted to the University of Edinburgh and in official positions and other ways did much to promote its advancement.

He wrote important books and articles on poisons and had the reputation of carrying out dangerous tests on himself. For much of this work he required a laboratory. Obituary notices of Christison do not state when or where this laboratory was established; Burdon

Sanderson in 1871 (*Nature*, *3*, 189) reports its existence and that it was open to any worker prepared to meet the current expenses of working there.

Sir Andrew Douglas Maclagan (1812–1900)

Andrew Douglas Maclagan was born at Ayr. His father, David Maclagan, was a Scottish gentleman and family doctor, who served as a physician through the Peninsular war; of his seven sons, three became doctors and the youngest Archbishop of York. Douglas entered the High School of Edinburgh in 1818 and thence passed to the University of Edinburgh to become MD in 1833. He had passed Licentiate of the Royal College of Surgeons of Edinburgh in 1831 and became Fellow in 1833. After graduation he studied further in London, Paris and Berlin before settling in Edinburgh with an appointment as assistant surgeon at the Royal Infirmary.

In 1845 he began to lecture on Materia Medica in the Extra-academical School, an appointment which he continued for eighteen years until, in 1862, he was appointed to the Chair of Medical Jurisprudence and Medical Police in Edinburgh; this chair he retained until he retired in 1896. He was regarded very highly as a responsible expert witness in court cases. Soon after his appointment he began to lecture in public health and this was added to the title of his chair. He established a Public Health Laboratory which was operating in 1871 (*Nature*, 1871, *3*, 189) although its full recognition as a laboratory of the University did not come until 1884. Maclagan was a leading authority on the analysis of poisons and made many contributions to toxicology and materia medica. In this laboratory Gamgee (p. 232) was assistant from 1863–9 and began his researches on haemoglobin; Brunton (p. 218) also worked there as a student for his thesis on digitalis.

Maclagan's position as Professor carried also an *ex officio* appointment as assistant in Clinical Medicine and so he lectured at the Infirmary, in later years particularly in dermatology, which he established as a special department. He became much respected and influential in the medical world of Edinburgh, President in turn of the Royal College of Surgeons of Edinburgh, The Royal Society of Edinburgh, and the Royal College of Physicians of Edinburgh. To have been President of both the Colleges of Surgeons and Physicians was almost a unique honour, the only other instance being his father before him. Maclagan was honoured by knighthood in 1886.

A feature of medical life in Edinburgh in the second half of the nineteenth century was the social clubs, such as the Royal Society Club, the Aesculapian Club, the Medico-Chirurgical Club. In this social life there was much interest in music, particularly singing, and both Christison and Maclagan are described in their obituary notices as having fine voices and, particularly Maclagan, as often singing works of his own composition. With this established social life, Edinburgh was in 1858 the very successful host of the annual meeting of the British Medical Association. The *British Medical Journal* (1858, p. 663−9) gives detailed reports of the speeches at the dinner and intersperses the comments: 'The proceedings were at this point varied by a glee, which was finely sung by Drs. Christison, Bennett, Peddie and Douglas Maclagan'. Later, 'Dr. Douglas Maclagan then sang a humorous original dittie, in praise of "Plain Cold Water", which evoked much merriment and applause'.

Glasgow

Andrew Buchanan (1798−1882)
Buchanan qualified MD Glasgow in 1822 and his whole working life was in that city. In his long career he was distinguished both as a surgeon and a physician. In his early days as a physician he fought energetically in the struggle with fever and cholera; as a surgeon he served in the Royal Infirmary and invented a rectangular staff for lithotomy, providing an operation which was used for many years, known by his name.

Buchanan was appointed Professor of Materia Medica in Anderson's College in 1824 and in 1839 Professor of the Institutes of Medicine (Theory of Physic). He retained this position for thirty-five years during which he played a full part in the affairs of the University and was for a period Vice-Rector. However his resignation from the chair in 1875 at the age of 77 was in an atmosphere of recrimination.

Although he regarded the duties of the chair as primarily clinical, he did some physiological work on the coagulation of blood and other body fluids and on the circulation, and taught such physiology as he could without equipment. The paucity of the teaching was revealed in 1875 by a report to the General Medical Council by the Visitors to Examinations, written by Humphry (p. 163) and Quain (p. 88). Buchanan took this as a personal attack and wrote an unwise letter to the *Lancet* (1875, i, 661) in which he attributed the failure to the lack of provision by the University. 'I ask Prof. Humphry if it be possible to teach practical physiology without instruments, without a laboratory, without funds and without an assistant'. The letter states that when he did receive some money for instruments, he could only buy such as he could carry to and fro from his house. Although a laboratory was then in the process of erection, there was no provision for its equipment or maintenance. In this atmosphere Buchanan resigned his chair; other reports imply that at the age of 77 it was high time.

He remained a respected member of the medical profession in Glasgow and was much gratified by being elected President of the Faculty of Physicians and Surgeons, fulfilling the duties of the post with grace and dignity. He was also honoured by the degree of LLD in 1881.

Buchanan was succeeded by M'Kendrick (p. 202). Although the medical school had greatly developed under the influence of Thomson, there was no provision for physiology; the subsequent development of physiology at Glasgow was due to M'Kendrick.

Allen Thomson, FRS (1809–84)

Allen Thomson was born in Edinburgh, the son of *John Thomson* (1765–1847), surgeon, physician and for a period Professor of Pathology in Edinburgh. John Thomson was also a writer, particularly known for his *Life of Cullen*, which was a major item in George Eliot's reading in preparation for the writing of Middlemarch. John Thomson was very well known to the great people of both Edinburgh and London and provided Sharpey an effective reference in his application for the chair at University College (p. 73).

Allen Thomson's education was directed by his father. He qualified MD Edinburgh and FRCS Edinburgh in 1830 and, encouraged by his father, determined to be an anatomist and physiologist. Like Sharpey, he first travelled in Europe, before in 1831 he joined Sharpey in lectures in an extra-academic school in Edinburgh (p. 78). Also in the next few years, Thomson visited London and was introduced by his father to important medical men such as Cooper and Henry Holland. With his father, Allen Thomson again visited many continental centres of anatomy, where he noted the best preparations and illustrations for teaching anatomy.

In 1839 he was appointed Professor of Anatomy in Marischal College, Aberdeen but in 1842 returned to Edinburgh as Professor of the Institutes of Medicine. Apparently his father had always intended that Allen should be Professor of Anatomy at Glasgow and, when this chair became vacant in 1848, Allen Thomson was elected to it, freeing the chair in Edinburgh for the appointment of Hughes Bennett (p. 97). For the next thirty years in Glasgow, Thomson was one of the makers of a great school of Medicine and was also greatly involved in the foundation of the modern University. Although Buchanan, Professor of the Institutes of Medicine, had become ineffective, when M'Kendrick succeeded to this chair in 1876, it was in a school of high standard in other respects. Thomson retired in 1877 and was succeeded by his assistant, Cleland.

After his retirement, Allen Thomson went to live in London in a house next door to his son, John Miller Thomson, who was then Demonstrator in Chemistry at King's College, London. Already a friend of Sharpey and well known to other physiologists, Allen Thomson was in 1878 elected a member of the Physiological Society and was honoured by election as Honorary Member in 1882. He died in London in 1884.

Allen Thomson was a great teacher of anatomy, with leanings towards physiology. His lectures attracted large numbers of students; they were carefully prepared and illustrated by diagrams and drawings with an elaborate synopsis written on the blackboard. About 1850 he contributed articles to Todd's *Cyclopaedia* (p. 36) and later became the editor of the Descriptive part of Quain's *Anatomy* (p. 28) in conjunction with Sharpey and later with Schäfer and Thane of University College. Thomson's research work was in descriptive embryology. His name is associated with no particular point but his general work is regarded as a background from which the subject was developed by Balfour (p. 247) and others. Allen Thomson's status was recognised by his election Fellow of the Royal Society of London in 1848. Later, after he had retired to London, Thomson was a member of the Council of the Royal Society and Vice-President. By far his greatest work in official positions was in Scotland. The obituary notice in the *Proceedings of the Royal Society* written by his former demonstrator, William Aitken, provides a detailed account of these positions and also those of his father. It sets out the great influence father and son had in medical education in Scotland.

Aberdeen

George Ogilvie (1820–86)
Ogilvie was born in Aberdeen and graduated AM in 1839. He then went to Edinburgh to qualify in medicine, MD Edinburgh in 1842

and Licentiate of the College of Surgeons, Edinburgh in the same year.

He soon returned to Aberdeen as Lecturer in the Institutes of Medicine in Marischal College. The status of medical education in Aberdeen was uncertain until in 1860 the 'fusion' of the two colleges was accomplished and Ogilvie was then given the position of Professor of the Institutes of Medicine in the University of Aberdeen in Marischal College. He held this post until his retirement in 1877. For a number of years he was secretary to the Faculty and in other ways also took part in the affairs of the newly-formed University.

Ogilvie was respected by students for his dignity, not unaccompanied by kindness, and was accepted as a good teacher particularly in his special subject of histology. He introduced microscopes; Ogilvie had been in Edinburgh at the time when Hughes Bennett was introducing the use of the microscope. However, Ogilvie's own interest was particularly in botany. He wrote *On the forms and structure of fern stems* (1859) and *The genetic cycle* (1861). He was a religious man and his thinking on natural science took a philosophical aspect in his best-known book *The master-builder's plan, or the principles of organic architecture as indicated in the typical forms of animals*, published in 1858 before Darwin's *Origin of species*.

In 1876 Ogilvie inherited the estate of Boyndlie, near Fraserburgh and adopted the name of Ogilvie-Forbes. He retired from his academic post to live the life of a country gentleman, taking an active part in the business of the county.

References

Coutts, J. (1909). *A History of the University of Glasgow from its foundation in 1451 to 1909*. Glasgow: Maclehouse.

Simpson, W.D. (1963). *The Fusion of 1860: a record of the Centenary Celebrations and a History of the United University of Aberdeen 1860–1960*. Edinburgh: Oliver and Boyd.

Turner, A.L. (1933). *History of the University of Edinburgh 1883–1933*. Edinburgh: Oliver and Boyd.

Biographies

Bennett: *British Medical Journal* (1875), ii, 473–8.
Buchanan: *Lancet* (1882), ii, 82–3.
Christison: *British Medical Journal* (1882), i, 214–5, 249–52.

Maclagan: *British Medical Journal* (1900), i, 935–7 (P).
Ogilvie: Simpson (1963), p. 254–5.
            British Medical Journal (1886), ii, 93.
Robertson: *British Medical Journal* (1909), i, 191–3 (P), 252.
Thomson: *Proc. Roy. Soc. London* (1887), **42**, xi–xxviii.

8

Physiologists not involved in medical teaching

Comparative anatomists

About 1860, comparative anatomy was an important subject and was taught in medical schools because it became a necessary subject for examinations at the College of Surgeons. Comparative anatomy was the scientific basis of the very wide discussion of the idea of evolution which dominated natural science after the publication by Darwin in 1859 of *The origin of species* and in 1871, *The descent of man*. Workers such as Darwin, Carpenter, Rolleston, Huxley were accurately describing animal species and so laying a sound foundation for future study of function. Several of these men had sufficient interest in physiology to associate themselves with the formation of the Physiological Society in 1876 and their influential position in general biology was recognised by their subsequent election as Honorary Members.

Charles Darwin (1809–82) was an acknowledged leader of British biologists and along with Sharpey, he was elected an Honorary Member at the foundation of the Society. This perhaps also recognised Darwin's evidence to the Royal Commission on Animal Experimentation and his contribution to the formulation of the Act of 1876. There is no record of his having attended meetings of the Society, but at this stage of his life, Darwin rarely left his house at Downe.

William Benjamin Carpenter, FRS (1813–85)
William Benjamin was born at Exeter, the son of Dr Lant Carpenter who was an Unitarian Minister, well known for his humane work. Later the family moved to Bristol, his father becoming master of an Unitarian School; William Benjamin was educated there and remained a staunch Unitarian throughout his life. He began medicine by apprenticeship to a doctor in Bristol, then took medical classes at University College, London and passed the examinations at the

College of Surgeons and Society of Apothecaries in 1835 and also became MD (Edinburgh) in 1839.

In 1839 he began practice as a physician in Bristol but in 1844 abandoned medicine to devote himself to comparative biology with particular interest, and later fame, in the comparative biology of marine animals. He went to London and in 1844 was appointed Lecturer in Medical Jurisprudence at University College and he also lectured in physiology at the London Hospital. He was a Professor at the Royal Institution and was soon elected Fellow of the Royal Society. From 1851–9 he was Principal of University Hall, the residential college of University College, London, and from 1856–79 he was Registrar of the University of London and, after his retirement, an active member of the Senate of the University of London.

Carpenter was an active lecturer to general audiences on biological subjects or microscopical science. For example in 1851 he gave a series of lectures in Manchester, staying with Mr and Mrs Gaskell (the novelist); Mr Gaskell as an Unitarian Minister was a natural host to Carpenter, also a staunch Unitarian. In 1855 on a visit to London, Mrs Gaskell left her card at Carpenter's home.

Scientifically, Carpenter became interested particularly in the structure and function of marine animals and marine biology in general. He initiated and reported the findings of scientific investigations on the voyages of 'Lightning' (1868), 'Porcupine' (1869–70), 'Shearwater' (1871) and 'Valorous' (1875). Much of his work was with the microscope and this led to the book *The Microscope and its revelations* which went through many editions. Like Huxley, he supported Darwin. He was one of the many zoologists associated with his friend and admirer, Ray Lankester, in the formation of the Marine Biological Association and the setting up of the laboratory at Plymouth (p. 246).

Carpenter was known to physiologists by his book *Principles of general and comparative physiology, intended as an introduction to the study of human physiology and as a guide to the philosophical pursuit of natural history*, first published in 1839. This book was recommended in 1840 by Sharpey (see p. 29) and was very successful. In 1842, Carpenter separated the book into two parts – one on comparative physiology and the other, *The principles of human physiology*, continued to be a popular text for medical students, until between 1864 and 1876 Power (p. 57) edited the seventh, eighth and ninth editions. The early editions dealt at length with ideas about mind and in 1874 this section was separated as *The principles of mental physiology, with their applications to the training and discipline of the mind and the study of its morbid conditions.*

Carpenter was well known and liked by the founders of the Physiological Society and was invited to become a founder member but declined the invitation, apparently because he no longer regarded himself as a physiologist; by this time Power had edited his book for over ten years. The high esteem in which Carpenter was held was shown by his election as an Honorary Member in 1882, together with Bowman, Huxley, Marshall, Paget and Allen Thomson.

George Rolleston, FRS (1829–81)

Rolleston was born at Maltby in Yorkshire. As a child he was precocious in the classics and in 1848 entered Pembroke College, Oxford, achieving in due course a first class degree in Classics. He then studied medicine at St Bartholomew's Hospital, obtaining the qualification of MRCP in 1856, and MD (Oxford) in 1857. He served in a civilian hospital at Smyrna through the Crimean war before in 1857 returning to London with the intention of practising medicine.

He was, however, almost immediately appointed to succeed Acland as Lee's Reader in Anatomy at Christchurch, Oxford and combined this post with medical practice, until in 1860 he was appointed Linacre Professor of Anatomy and Physiology in the University and became a fellow of Merton College. He then abandoned medical practice to devote his time to his University post, which he held until his death in 1881. He worked hard in the administration of the University, being for many years a member of the Council of the University and its representative on the General Medical Council. He was much respected as a teacher and in University and scientific circles.

The duties of the Linacre Professor involved teaching over a wide range of subjects, including human anatomy, comparative anatomy, anthropology and physiology. Rolleston often stated the impossibility that one man could be expert in so many fields and saw the need for separate chairs. It was only after his death that there was a separate Waynflete Professorship of Physiology, first occupied by Burdon Sanderson (p. 141).

Rolleston concentrated his own teaching on comparative anatomy and instituted practical teaching in this subject. He set up a carefully selected series of animals – rat, pigeon, frog, perch, crayfish, beetle, anodon, leech and tapeworm. He had mounted dissections of these and students worked through the series, also dissecting for themselves. This was then becoming the way of teaching comparative anatomy, also used by Huxley and further developed by Ray Lankester (p. 245), who was after 1866 Rolleston's student and admirer. Parallel to this form of teaching Rolleston was much involved in the development of the zoological section of the University Museum, which was built and opened in 1860. Rolleston caused the collection at Christchurch, for which he was responsible as Lee's Reader, to be transferred to the University Museum to add to other collections. His organisation of this section was further developed by Lankester. Rolleston's notes for the students' practical classes became the basis of a book *Forms of animal life* (1870) which was very widely read.

In research, Rolleston's work was particularly in anthropology, in which he was an accepted authority; he was elected FRS in 1862. His historical bent produced articles such as 'Domestic pig in prehistoric times' and 'Domestic cats through the ages'.

Although Rolleston was Professor of Anatomy and Physiology, he made no contribution to the development of physiology as an experimental subject. In 1868 in his Address to the Section of Physiology at the annual meeting of the British Medical Association (*BMJ*, 1868, ii, 155), Rolleston attempted to relate medicine to the various aspects of his own chair, physiology, human and microscopic anatomy, comparative anatomy. The section on physiology contains little understanding of experiments and measurement, in this respect far behind Sharpey's similar Address of 1862. Sharpey had then welcomed the recent appointment of a Professor of Physiology at Oxford, but clearly Rolleston's work did not fulfil what Sharpey had hoped. Rolleston was one of those asked to join in founding the Physiological Society in 1876; in the records of the Society, Rolleston's name carries the terse comment 'refused to join'. For all his qualities in other fields, Rolleston was not an experimenter and it was perhaps a legacy from his reign that in Oxford Burdon Sanderson in 1883 was confronted by antivivisectionist opposition (p. 144). In his own field of natural science Rolleston was a great figure and modern zoologists can well regard him as one of their founders.

Thomas Henry Huxley (1825–95)

Huxley was born at Ealing, the son of a schoolmaster. With no other career apparent, he began to study medicine as apprentice to a brother-in-law and in 1842, Thomas Henry and his brother James entered

Charing Cross Medical School together. Thomas Henry was a successful student and was much influenced by the lectures of Wharton Jones (p. 54). In the course of instruction on the use of the microscope, Huxley observed a layer of cells in hair follicles, known as Huxley's layer.

When he qualified with the Diploma of the College of Surgeons in 1846, Huxley was employed as a surgeon by the Royal Navy and was appointed assistant surgeon in *HMS Rattlesnake*, about to begin a voyage surveying in the Torres Straits, north of Australia. There was already a naturalist and a surgeon appointed to the ship, so that Huxley was a supernumerary able to follow his own scientific interests. Particularly he collected and studied marine animals. The voyage of the *Rattlesnake* lasted four years and, back in London in 1850, Huxley retained his position as a naval surgeon but was allowed to live in London working on the material he had collected on the voyage. Papers sent from the *Rattlesnake* to the Linnean and Royal Societies were so well received that he was elected FRS in 1851 and awarded a Royal Medal in 1852. In 1853, however, the Navy insisted that he undertook surgical work and Huxley resigned his naval appointment. Now without a paid position, he had come to realise that he wanted to be a physiologist and unsuccessfully applied for posts in Toronto and as Professor of Anatomy and Physiology at King's College, London, where Beale (p. 43) was given preference over him.

In 1854, he was appointed Lecturer in General Natural History at the Metropolitan School of Science applied to Mining and the Arts (School of Mines), then in Jermyn Street. He refused a parallel post of Palaeontologist, saying 'I did not care for fossils and ... should give up Natural History as soon as I could get a physiological post'. In fact Huxley remained Lecturer in Natural History for thirty years and a large part of his work became palaeontological, as he was also appointed scientist to the Geological Survey of Great Britain.

The guaranteed salary of £300 per annum allowed him to marry Henrietta Heathon, whom he had met in Sydney in 1847 and who could at last come to England to be married in 1855. His income, however, had to be supplemented by fees for 'articles written for any editor who was prepared to give adequate pay to a pen able to deal with scientific themes in a manner at once exact and popular, incisive and correct' (Foster). Huxley also gave many popular lectures for which fees were paid, some at the School of Mines and the Royal Institution.

In November 1859, Darwin's *Origin of species* was published. Huxley had met Darwin in 1851–2 and he was one of several friends who had known of Darwin's argument before it was published. Immediately Huxley became the expounder of Darwinism to all sorts of audience. Huxley saw the doctrine of natural selection as eminently suitable as a topic for all audiences and his support of the doctrine made him well known outside the immediate world of science. From 1860, Huxley was spreading the doctrine of biological science to an ever-widening audience. Huxley had a deep conviction 'that science was not merely for men of science alone, but for all the world: and that not in respect of its material benefits alone, but also, and even more, for its intellectual good' (Foster). It was for this reason that in 1870, Huxley allowed himself to be elected to the London School Board and thereafter he had great influence in introducing science into schools.

In addition to his work as a writer and lecturer, Huxley prosecuted his scientific investigations, chiefly the continuation of morphological work on sea and other animals, extending into the comparative morphology of vertebrates. Apart from his other work, Huxley was a great and original zoologist. There was also work in physiological histology, producing papers on teeth and on *Corpuscula Tactus* in the *Quarterly journal of microscopical science* and a large contribution on tegumentary organs in Todd's *Cyclopaedia* (p. 36).

Both by his success as a popular expositor and also as a true scientist, Huxley became an accepted authority in natural science. He was examiner in physiology and comparative anatomy in the University of London (1858–70). In 1871–80 he succeeded Sharpey as secretary of the Royal Society and in 1883–5 was President. He was appointed to many official posts and commissions, one of which, 'The Royal Commission on the Practice of subjecting Live Animals to Experiments for Scientific Purposes (1875–6)', was of great importance to physiologists. The growing science of physiology had in Huxley a most powerful support. He certainly had an influence in inducing Trinity College, Cambridge, to appoint Michael Foster as Praelector in Physiology in 1870 and he influenced Jodrell to endow the Chair of Physiology at University College. Huxley was a founder of the Physiological Society and in 1882 was elected an Honorary Member.

Huxley made two important contributions to the teaching of physiology. One was his book *Lessons in elementary physiology*

(1866), based on a course of lectures to school teachers in 1865
(p. 31). For many years this was the standard elementary book
for general students and was also an introductory book for medical
students. Secondly in 1872 when the School of Mines moved to
South Kensington, and laboratory space became available, Huxley
introduced classes in biology which included practical work. Huxley
would give a lecture early in the afternoon and the students then
formed groups of about eight in the laboratory, where by dissection
and microscopic work under a demonstrator, they actually saw
the material relevant to the lecture. Among his demonstrators
were Lankester and Romanes who became zoologists, Thiselton-
Dyer and Vines who became botanists and Foster, Rutherford,
Newell Martin and Ferrier who became physiologists of the late
1870s. Particularly in the case of Foster, this association had great
results in the development of practical class teaching of physiology
in Cambridge and in making physiology as one part of general
biology.

After 1880, Huxley became less active as a zoologist and more of
a philosopher. His influence was still great but of less specific
importance to physiology. He died in 1895.

Writers and philosophers

In his address on physiology in 1862 (p. 85), Sharpey said that physiology
was coming to the attention of a wider audience; Huxley by his lectures
and his textbook introduced physiological ideas to audiences of teachers,
working people and others. G. H. Lewes similarly introduced physiology
to the reading public. He was well known in the intellectual life of London
and after about 1860 devoted himself to work and writing on biology
and physiology; both Lewes and Huxley were founder members of the
Physiological Society in 1876, with Lewes acting as secretary at the meetings
by which the Society was founded. Like Huxley, he was in 1876 an established
senior member in the intellectual activities in London and the influence of
these men was of great importance.

In considering G. H. Lewes as a physiologist, it is necessary to consider
particularly one group in the intellectual life of London at that time.
Herbert Spencer, 'George Eliot' and Lewes were prominent members
of a group of advanced thinkers, whose philosophical thinking was expressed
by the theoretical writings of Spencer.

Herbert Spencer (1820–1903)

Spencer's work is regarded as sufficiently important in the history of philosophy to be allowed two pages in the modern *Encyclopaedia Britannica*. Certainly he was highly regarded by his own group. In 1854 George Eliot wrote: 'he will stand in the Biographical Dictionaries of 1954 as an original and profound philosophical writer, especially known by his great work which gave a new impulse in psychology and had mainly contributed to the present advanced position of that science'.

Born in Derby of a family with strong connections with Wesley, Herbert Spencer showed independence in thought through his years at school. When about sixteen years of age, he obtained an appointment as assistant to the engineer of the London and Birmingham Railway and opened a promising career as an engineer in the building of railways. However, partly by mischance and partly by choice, he entered the field of social philosophy, being sub-editor of a provincial paper before becoming in 1848 sub-editor of the *Economist* in London. Thereafter he was writing articles on philosophical subjects, which like all his subsequent work, consisted of his own ideas without much reading of existing literature. In 1860 he issued a 'Programme of a system of synthetic philosophy', under which he produced from 1862 onwards a series of articles collected into books such as *Principles of biology* (1864–7) and *Principles of psychology* (1870–2). As early as 1850 his work, on purely philosophical grounds, was leading to ideas of evolution as the method of development of social systems and when in 1859 Darwin published *Origin of species*, Spencer became a strong supporter of Darwinism. Already in 1852, in an article 'A theory of population' in the *Westminster review*, Spencer had coined the phrase 'the survival of the fittest'. Similarly in his writings on sociology, Spencer had pointed out that a social organisation could survive only if it adapted to changing surroundings and he defined *adaptation* as the 'continual adjustment of internal relations to external relations'. Later physiologists, such as Starling (1912) accepted this definition and the idea of adaptation was linked with 'constancy of the internal environment' before the term *homeostasis* was introduced by Cannon in 1929.

Spencer had considerable influence but we cannot rate him as a physiologist because he dealt solely in philosophical thinking, not facts. Huxley (who was a friend) said that 'Spencer's definition of a tragedy was the spectacle of a deduction killed by a fact'.

Spencer and Mary Ann Evans (George Eliot) became such close friends that it was thought they might marry; they spent many evenings walking together on the Savoy terraces, discussing philosophy. According to letters to friends, they discussed their relationship, decided that they were not in love and therefore there was no reason why they should not have as much of each other's society as they liked. Spencer was also a friend of G. H. Lewes and introduced him to Miss Evans. After 1854 Lewes and Miss Evans lived as man and wife in an unconventional but very fruitful union, which transformed Mary Ann Evans into George Eliot and G. H. Lewes into a physiologist.

Mary Ann Evans (1819–80): 'George Eliot' (1857–80)

'George Eliot' was born Mary Ann Evans at South Farm, Arbury on 22 November 1819. Her father was agent and manager of the Arbury estate of the Newdigate family, who were benefactors of Oxford University, founding the Newdigate Prize for poetry. Later her family moved to Griff House on the Arbury estate and the country scenes of many of George Eliot's novels are based on her girlhood memories of this area between Nuneaton and Coventry. She went to school in Nuneaton. Her mother died in 1836 and her sister married, leaving Mary Ann to keep house for her father and brother at Griff House, until her father retired in 1841, when he and his daughter moved to Foleshill Road on the outskirts of Coventry. Her father died in 1849, after a long illness in which he was cared for by Mary Ann.

After the death of her father, Mary Ann spent nearly a year in Geneva, until in 1851 she began working in London as assistant editor of the *Westminster Review*. The editor was John Chapman whose house was the meeting place of advanced philosophical thinkers. Marian Evans (as she now called herself) was well qualified for this position. In Coventry she had become friendly with Charles Bray and his wife. Bray was a well-to-do ribbon manufacturer but also a writer of books on philosophy and through him Marian Evans had been commissioned to produce an English translation of Strauss's *Life of Christ*. The result was a critical examination of the Christian religion, extremely advanced for its time and certainly sufficient to draw the attention of philosophers to her capabilities.

Thus she entered the group which included Spencer and Lewes. Her first impressions of Lewes were not altogether favourable, but they grew together; in April 1853 she wrote to Mrs Bray: 'People are very good to me. Mr Lewes especially is kind and attentive, and has quite won my regard, after having had a good deal of my vituperation. Like a few other people in the world, he is much better than he seems. A man of heart and conscience wearing a mask of flippancy'. A year later, after serious thought, Lewes and Miss Evans formed their union for life, signalled by their departure together on 20 July 1854 for a year in Weimar and Berlin. Thereafter she always signed herself Marian Evans Lewes and their irregular union came to be generally accepted in London, although she was rejected by her brother.

In 1855 both were doing very similar work. Both were writing major books of a philosophical nature, *Life of Goethe* by Lewes and a translation of Spinoza's *Ethics* by his wife. Also both were working very hard in editing and writing for periodicals. They needed money to support and educate Lewes's three sons. Marian was accepted as their mother and accepted her responsibility for them.

In 1857, they discovered that Marian could write novels and Lewes was so well known as a literary critic that he could present them to Blackwood for publication as the work of a friend, George Eliot, who wished to remain anonymous. *Scenes of Clerical Life* (1858), *Adam Bede* (1859), *The Mill on the Floss* (1860) and *Silas Marner* (1861) were highly successful and after 1860 Lewes and George Eliot were free of financial worry and able to write what they wanted to write. They no longer needed to write articles for magazines. George Eliot thereafter wrote more philosophical and carefully studied novels – *Romola* (1865), *Felix Holt* (1866), *Middlemarch* (1872) and *Daniel Deronda* (1875) – of which only *Middlemarch* is now much known. Works other than novels – *Spanish gipsy* (1868), *Legend of Jubal* and other poems (1874), *Armgart* (1870) are now virtually forgotten except to students writing theses.

Lewes died on 28 November 1878. Three volumes of his *Problems of life and mind* had been published; the fourth volume was prepared for the press by George Eliot assisted by Michael Foster and was published in 1879. In memory of her husband, George Eliot founded the George Henry Lewes Studentship, helped by friends including Foster (p. 167). She wrote to a friend explaining her purpose: 'The studentship is to supply an income to a young man who is qualified and eager to carry on physiological research, and would not otherwise have the means of doing so. I have been determined in my choice of the studentship by the idea of what would be a sort of prolongation of *his* life. That there should always, in consequence of his having lived, be a young man working in the way he would have liked to work, is a memorial of him that comes nearest my feeling'. The studentship was advertised in September 1879:

George Henry Lewes Studentship – This studentship had been founded in memory of Mr George Henry Lewes, for the purpose of enabling the holder for the time being to devote himself wholly to the prosecution of original research in physiology. The Studentship, the value of which is slightly under £200 per annum, paid quarterly in advance, is tenable for three years, during which time the student is required to carry on, under the guidance of a director, physiological investigations to the complete exclusion of all other professional occupations. No person will be elected as a 'George Henry Lewes Student' who does not satisfy the Trustees and Director, first, as to the promise of success in physiological inquiry; and second, as to the need of pecuniary assistance. Otherwise all persons of both sexes are eligible. Applications, together with such information concerning ability and circumstances as the candidate may think proper, should be sent to the present Director, Dr Michael Foster, New Museums, Cambridge, not later than October 15, 1879. The appointment will be made and duly advertised as soon as possible after that date.

The studentship has fulfilled its founder's wishes by supporting many who became famous physiologists – Sherrington, A. V. Hill, Dale, Mellanby and many others. On the other hand some of George Eliot's ideas have not been fulfilled. She was interested in the advancement of higher education for women and was associated with the foundation of both Newnham and Girton, but there have been few women holders of the studentship. Secondly, George Eliot did not specifically associate the studentship with Cambridge – 'It is to be at Cambridge to begin with, and we thought at first of affiliating it to the University: but now the notion is that it will be well to keep it free, so that the trustees may move it when and where they will'. Michael Foster was the obvious first director and hence its original association with Trinity College, Cambridge, where it has in practice largely stayed. The first student was C. S. Roy (p. 183), who was nominated by Foster at a meeting of the trustees on 18 October 1879.

George Eliot greatly missed the close companionship and encouragement that Lewes had supplied. After his death she attempted no serious writing; *The Impressions of Theophrastus Such*, which was published in 1879, had in fact been written in 1878. Towards the end of 1879 and in the early months of 1880, George Eliot was often accompanied by a much younger man, J. W. Cross and in April 1880 they married. After a trip of about three months to Venice, they returned to London in July 1880. George Eliot was then frequently ill and died on 22 December 1880.

Cross later published a biography, *George Eliot's Life, as related in her letters and journals*. This is almost entirely composed from George Eliot's own writings and is indeed a great biography, comparable to Lockhart's *Life of Scott* and Mrs Gaskell's *Life of Charlotte Bronte*. However, Cross's interests were literature and philosophy and he apparently had no interest in physiology. Comparison of his extracts from George Eliot's letters with the more complete letters edited by Haight (1955) shows that Cross has omitted some of the relatively few references to Lewes's activity in physiology. Such references are in any case few since George Eliot was usually writing to her Coventry and other literary friends or to her publisher, Blackwood. There is enough to show that George Eliot knew and appreciated what Lewes was doing but had no active part in it. Their relationship was one of close companionship and mutual encouragement but they did not collaborate in any piece of work, literary or scientific.

George Henry Lewes (1817–78)

G. H. Lewes was born in London, the son of actors and the grandson of Charles Lee Lewes (1740–1803), a famous actor. George Henry's education was desultory, in London, Jersey and Brittany and apparently for long periods on the continent, by means of which he was fluent in French and German. Little is known of his early life. He considered becoming a doctor and walked the wards of St George's Hospital long enough to dislike what he saw.

In London after 1840 he acted a little professionally and played in Dickens's amateur company. Soon, however, he was earning his living by freelance writing for periodicals on a wide range of subjects; it was said that he would write on anything an editor would pay him for. Examples of articles in periodicals included the French drama, Shelley, the errors and abuses of English criticism, the circulation of the blood and so on. He was a well regarded literary critic and reviewed *Jane Eyre* and *Shirley*, thereby coming into half friendship and half quarrel with Charlotte Brontë because, although he praised the books, he did not ignore the fact that they were written by a woman.

By 1850 he was well established in the London literary world and in the group of advanced philosophy and when in 1850 Thornton Leigh Hunt established the *Leader*, Lewes became co-editor dealing particularly with literary matters. Advanced thinking showed its disadvantages when Lewes's wife had two children by Thornton Leigh Hunt. In 1854 Lewes was separated from his wife but could not obtain a divorce because he had condoned the first offence.

Lewes's literary and scientific career can be seen by the following list of his published books, a total of fifteen in thirty years, in addition to his work as a journalist.

> *Biographical history of philosophy* (1845–6).
> *Ranthorpe* (a novel, 1847).
> *The Spanish drama*; Lopa de Vega and Calderon (1847).
> *Rose, Blanche and Violet* (a novel, 1848).
> *The noble heart* (a play, 1848).
> *Life of Maximilien Robespierre* (1849).
> *Comte's Philosophy of the sciences* (1853).
> *The Life of Goethe* (1855).
> *Seaside Studies at Ilfracombe, Tenby, Scilly Isles and Jersey* (1858).
> *Physiology of common life* (1859–60).
> *Studies of animal life* (1862).
> *Aristotle, a chapter from the history of the sciences* (1864).
> *Problems of life and mind* (1874–9).
> *On actors and the art of acting* (1875).
> *The study of psychology; its object, scope and method* (1879).

The years 1854–1857 were a turning point in Lewes's life. After 1854 he had the companionship and support of George Eliot; he was supporting her work and he himself changed to work on biology. Lewes's writing on literary topics culminated in *The Life of Goethe*, which was the first biography of Goethe in English and remained a standard work for many years. It was almost complete before their 'marriage' and was completed during the year of their residence in Germany in 1855. *Comte's Philosophy of the sciences* (1853) contained enough science and enough discussion leading towards the idea of evolution to bring Lewes into scientific conflict with Huxley, whose criticisms Lewes could repel by his superior knowledge of current French literature. In the controversy, Huxley accused Lewes of being only a book scientist, which produced the comment from Lewes that for some fifteen years he had been conducting experiments in physiology, which means that the had been working in biology during the years of his intense journalistic activity. In 1860, Lewes bought a new Smith and Beck microscope to replace one he had used for many years.

Serious writing in pure biology began with *Seaside Studies*. Lewes and Marian Evans, now his 'wife', went to Ilfracombe on 8 May 1856, 'with our hamper of glass jars, which we meant for our seaside vivarium'. They greatly enjoyed the open air and country walks but the object was to collect and study, particularly sea anemones. Marian wrote 'Every day I gleaned some little bit of naturalistic experience, either through G's calling on me to look through the microscope, or from hunting on the rocks'. In another letter to Mrs Bray, she says, 'You would laugh to see our room decked with yellow pie dishes, a foot-pan, glass jars and phials, all full of zoophytes, or molluscs, or annelids – and still more, to see the eager interest with which we rush to our 'preserves' in the morning to see if there has been any mortality among them during the night. ... We have made the acquaintance of a charming little zoological curate here (Mr Tugwell), who is a delightful companion on expeditions and is most good-natured in lending and giving apparatus and "critturs" of all sorts'.

On 26 June they moved to Tenby 'for the sake of making acquaintance with its molluscs and medusae'. On 9 August they returned to their home at Richmond, Surrey.

The visit to Ilfracombe and Tenby marks the beginning of Lewes's career as a biologist. It was also the beginning of George Eliot as a novelist. Lewes had often advocated that 'she must try and write a story'. Nothing came of it until one day at Tenby the idea of *The Sad Fortunes of the Reverend Amos Barton* drifted into her mind. Some journalistic work had to be finished first but on 22 September, after they had returned to Richmond, she began to write this story; it was finished on 5 November and sent by Lewes to the publisher Blackwood as the first of a proposed series. His letter began 'I trouble you with a MS of *Sketches of Clerical Life*, which was submitted to me by a friend who desires my good offices with you'. Blackwood willingly accepted the story and it was published in *Blackwood's Magazine* of January 1857, after some discussion of future plans, during which Marian adopted the name of George Eliot; George because it was Lewes's Christian name and Eliot 'because it was a good mouth-filling, easily pronounced word'. The second Scene of Clerical Life, *Mr Gilfil's Love Story* was begun on Christmas Day, 1856.

Seaside studies were resumed by travelling to the Scilly Isles on 17 March 1857 and *Mr Gilfil's Love Story* was finished and *Janet's Repentance* begun on this trip. They continued zoological collecting and on 12 May went on to Jersey. From there she wrote 'Our little lodgings are very snug – only 13*s* a week – a nice little sitting room, with a work room adjoining for Mr Lewes, who is at this moment in all the bliss of having discovered a parasitic worm in a cuttle fish'. The letters from Jersey draw the picture of their life together, working independently day by day, taking an interest in what the other did. At night they read together: 'The *Life of George Stephenson* has been real profit and pleasure. I have read Draper's *Physiology* aloud for grave evening hours and such books as Currer Bell's *Professor*, Mlle. de Auny's *Mariage en Province* and Miss Ferrier's *Marriage* for lighter food'.

They returned to Richmond on 24 July, *Janet's Repentance* was finished on 9 October and *Adam Bede* begun on 22 October 1857.

From this time they both had defined careers to follow. The anonymous George Eliot was a recognised writer whose novels sold extremely well. When *Scenes of Clerical Life* was issued as a book in January 1855, copies were sent to Froude, Dickens, Thackeray, Tennyson, Ruskin and Faraday who all replied with great appreciation of the author. Marian Evans, the struggling and uncertain journalist had become George Eliot, successful novelist, confident of her abilities and in the support of Lewes. Lewes had become an accepted biologist. His *Seaside Studies* appeared in *Blackwoods Magazine* during 1857–8 and as a book in 1858. A copy was sent to Huxley, who expressed his thanks and appreciation and no longer dismissed Lewes as a book scientist; thereafter they were on friendly terms.

Seaside Studies was also appreciated in Europe. When in January–March 1858 Mr and Mrs Lewes visited Munich and other cities, they were received by both literary and scientific communities. Thus, of their visit to Munich, George Eliot wrote 'Mr Lewes is in a state of perfect bliss this morning. He is gone to the Akademie to see wonders through Siebold's microscope and watch him dissecting'. In the evning they were at Siebold's house and saw 'the prettiest little picture of married life – the great comparative anatomist (Siebold) seated at the piano in his spectacles playing the difficult accompaniments to Schubert's songs, while his little round-faced wife sang them with much taste and feeling'. Lewes also worked with Bischoff and Harless: 'I have been doing some capital work in the laboratories of my friends here, where an extensive apparatus and no end of frogs are at my disposal! When government establishes a physiological institute professors (and amateurs) can work in clover'. Lewes also saw Jacubowitch's 50,000 preparations on the spinal cord with no evidence of fibres connected with cells. They met Liebig, of whom George Eliot wrote 'His manners are charming – easy, graceful, benignant, and all the more conspicuous because he is so quiet and low-spoken among the loud talkers here. He looks best in his laboratory, with his velvet cap on, holding little phials in his hand, and talking of Kreatine and Kreatinine in the same easy way that well bred ladies talk scandal. ... It was touching to see his hands, the nails black from the roots, the skin all grimed'.

Munich at that time was a great intellectual centre, with arts and science equally supported by the King of Bavaria, Maximillian II. The king himself was a scholar and every Saturday evening held a 'Tafelrund' to which scientists and writers were commanded and of which Liebig was always a member.

Although Lewes had become recognised as a biologist, he remained a writer and *Physiology of common life* and *Animal studies* were intermingled with other work. These physiological books were first published, like his other literary work, as serial numbers in magazines and then collected into books; they were written to appeal to a wide range of readers. In October 1859, George Eliot wrote to Lewes's son Charles 'Your father's *Physiology of common life* is selling remarkably well, being much in demand by medical students ... There is to be a new edition of *Seaside Studies* at Christmas – a proof·that this book also meets with a good number of readers'. Their income was mostly from their books; rough estimates are that in 1859 Lewes received about £500 from these two physiological books and George Eliot about £1,500 from *Clerical Lives* and *Adam Bede*. This comparative affluence allowed them to take larger houses, at Wandsworth in 1859, Blandford Square in 1860 and finally in 1863, the Priory in St John's Wood. Without any particular wish on their part, other than the pleasure of being visited by friends, the Priory became a social centre where they received eminent scientists and literary figures. Lewes was the effective host of any large gathering – a brilliant talker, a delightful raconteur, versatile, full of resource in the difficulties of amalgamating diverse groups (Cross, 1902). When any good performer happened to be present, there was music. George Eliot herself did not like large groups; her preference was to converse quietly with an interested friend, with her body bent forward to get as close as possible to the person with whom she talked. Of the two, it was George Eliot who left the deepest impression. When James

Paget said of her that 'she was the greatest genius — male or female — that we can boast of', he was expressing the opinion of many people of the 1870s.

The role of Lewes and George Eliot in the development of biology, physiology, philosophy, psychology is difficult to assess. They had no official positions, no membership of the Royal Society, no offices in the British Association, no contact with the University of London. Neither of them gave lectures either as part of a course like the medical physiologists nor to general audiences like Huxley. Yet historians, such as Trevelyan (1944), could say that her novels were taken by many as restating the moral law, in terms acceptable to the rationalistic agnostic conscience. Lewes introduced a physiological bias into the philosophy of the later Victorian literature.

Lewes cannot be regarded as a founder of experimental physiology; after 1870 that was the work of Burdon Sanderson and Foster. *Seaside Studies* and *Studies in Animal Life* are mainly comparative zoology, and the manner of collection of data has been described above. Writing in 1859 to Lewes's son, George Eliot said: 'I wish you could have seen today, as I did, the delicate spinal cord of a dragon-fly — like a tiny thread with tiny beads on it — which your father had just dissected! He is so wonderfully clever now at the dissection of these delicate things, and has attained this cleverness entirely by devoted practice during the last three years'. Presumably he had a laboratory at his house, although I have met no direct reference to it. Such experimental work as Lewes did, seems to have been on nerve and muscle in frogs and was in fact rather remote from the main interest of the later part of his life. Perhaps influenced by Spencer, he thought along lines we would now call psychology and *Problems of life and mind* must be regarded as a book contributing towards modern psychology rather than neurophysiology. By regarding the mind as a physiological problem, Lewes did perhaps foreshadow Sherrington's philosophical books.

Sir Francis Galton, FRS (1822–1911)

Francis Galton was born in Birmingham. His father was a prosperous business man; his mother was the daughter of Erasmus Darwin by a second wife, making Francis Galton a half cousin of Charles Darwin. Educated at King Edward's School, in 1838, aged sixteen, he travelled as far as Vienna with two young medical students, one of whom was Bowman (p. 38). Intended for the medical profession,

Galton was apprenticed to doctors in Birmingham before entering King's College, London in 1839. In London he lived in the house of Partridge (Professor of Anatomy), was taught physiology and anatomy by Bowman and Todd (p. 36) and did quite well in the first year. Then he set out on the continent with the official purpose of visiting medical centres. He extended his travel to Constantinople and Smyrna, because he just liked such travelling. Then he entered Trinity College, Cambridge to read mainly mathematics and took an ordinary degree in 1844. He had become something of the man about town and, also, ill health prevented him from seeking an honours degree.

In 1844 Galton was made well-to-do by the death of his father and abandoned medicine, for which he had no interest. He partly lived the life of a sporting gentleman but also could now exercise his love of travel. In 1845 he travelled up the Nile as far as Khartum and into Syria. He became a member of the Royal Geographical Society of which he was later president for many years. From this contact he formed the wish to make at his own expense a true journey of exploration into tropical South Africa. From Walfish Bay he struck inland into Damaraland, then completely unknown, now a district of South-West Africa. Galton's exploration was recognised by medals of Geographical Societies and as an explorer he was elected FRS in 1856. He described his exploration in a book *Tropical South Africa* (1853) and he also wrote a valuable vade-mecum for travellers *The art of travel* (1855). His health was so damaged by the South African journey that he was unable to make further explorations.

Galton married in 1853 and settled in London where he went much into society in literary and scientific circles. The home at 42 Rutland Gate, Hyde Park became a social centre where Mr and Mrs Galton were popular and charming hosts. Galton took a full share in the administration of scientific activities, although he held no University position. For example, in the British Association for the Advancement of Science, he was General Secretary from 1863–7, four times a Sectional President and twice refused to be President on account of his deafness.

From this assured financial and social position Galton made major contributions to a wide range of practical science.

In 1863 he published *Meteorographica or methods of mapping the weather*, in which he brought weather forecasting to its modern basis; he first used the term 'anticyclone'. As a member of the Meteorological Committee, Galton was closely associated with the first attempts to organize a meteorological service and the development of Kew Observatory.

Galton collected, classifed and arranged fingerprints to demonstrate that the fingerprints of an individual form an unique and permanent characteristic. He thus confirmed existing suggestions that the fingerprints could be used in criminology.

The great mass of his work was connected with heredity in man. In beginning this work he perceived that human characteristics had only been described without any attempt at quantitative measurement or study. In 1864 at the International Health Exhibition, Galton initiated an Anthropometric Laboratory where statistics were collected from many people concerning acuteness of senses, strength, weight, dimensions. He invented methods of making such measurements and sought methods for their statistical expression. This was the forerunner of a Biometric Laboratory subsequently established at University College, London. Galton sought a method of making composite photographs to show the facial characteristics in geographical areas or in families. He tried to apply similar methods to define the psychological characteristics of individuals and groups.

Galton was on friendly terms with his cousin Charles Darwin and, when *Origin of species* was published in 1859, became a strong supporter of the idea of evolution. His reflections on its possible significance in human heredity led to studies of whether those who obtained distinctions in Cambridge were related to those similarly honoured in other generations. This and similar studies and thinking resulted in a series of books, e.g., *Hereditary genius* (1869), *Natural inheritance* (1889), and *Noteworthy families* (1906). Appreciating that statistics were involved, Galton inspired mathematicians, especially his friend Karl Pearson (1857–1936) to formulate new methods. Galton became convinced that all attributes may be heritable and so proceeded to think of the improvement which might ultimately be produced in the human race by selective breeding. For this he used the term 'Eugenics'. To promote investigation, Galton initiated a Eugenics Laboratory to be associated with the Biometrical Laboratory already at University College, with Pearson as the active worker and he also endowed a research fellowship and scholarship; a journal *Biometrika* was established. Finally, in his will, Galton left the residue of his estate, £45,000 to endow a Chair of Eugenics at University College, London, expressing the wish that Karl Pearson should be the first Professor.

Formal public recognition of Galton's work came late in his life. The Royal Society awarded him medals in 1886, 1902 and 1910, the Anthropological Institute in 1901 and the Linnaean Society in 1908. He was knighted in 1909. He published an autobiography, *Memoirs of my life* in 1908 and after his death an extensive biography was written by Karl Pearson, *The life, letters and labours of Francis Galton* (3 vols., 1914).

Galton was an ingenious inventor, particularly of means to make measurements on human subjects; the Galton whistle was used to define the pitch threshold of auditory perception. An amusing example was the equipment he carried when going to see a public spectacle, for which purpose he designed a periscope – or he might carry a wooden brick wrapped in paper, with a long string attached, by which the brick was unobtrusively lowered, used as a pedestal and equally unobtrusively recovered.

Galton made no specific contribution to physiology, although in the general sense of measurement and quantitative analysis he was leading a movement from descriptive qualitative science towards the true science of quantitative measurement. In 1876 when the Physiological Society was formed, Galton was a major figure in the scientific and literary world of London and his support was valuable. Galton was present at the initial meeting at Burdon Sanderson's house (p. 262) and was one of the committee to draw up the constitution of the Society. He attended the first meeting on 26 May 1876 and continued to attend meetings, being on several occasions chairman, until his resignation in 1882. Sharpey-Schafer (1927) describes him as a cultured, modest and delightful man, always attentive and encouraging in talk about any scientific topic.

References

Cross, J. W. (1902). *George Eliot's Life, as related in her letters and journals. Arranged and edited by her husband.* 2 vols. (P). Edinburgh and London: Blackwood.
Darwin, F. (1892). *Charles Darwin: his life told in an autobiographical chapter and in a selected series of his published letters.* Edited by his son (P). London: Murray.
Haight, G. S. (ed.) (1954). *The George Eliot Letters.* London: Oxford University Press.
Huxley, L. (1900). *Life and Letters of Thomas Henry Huxley.* London: MacMillan.

Pearson, K. (1914–30). *The Life, Letters and Labours of Francis Galton.* 3 vols. (P). Cambridge University Press.

Sharpey-Schafer, E. (1927). *History of the Physiological Society during its first Fifty Years, 1876–1926.* Issued by the Society and published as a supplement to *Journal of Physiology*, December. Cambridge University Press.

Smith, R. E. (1960). 'George Henry Lewes and his "physiology of common life", 1859', *Proc. Roy. Soc. Med.*, **53**, 569–74 (P).

Trevelyan, G. M. (1944). *English Social History*, pp. 563–4. London: Longmans Green.

Biographies

Carpenter: *Proc. Roy. Soc. London* (1886), **41**, ii–ix.
              Dictionary of National Biography, **9**, 166–8.
Darwin: *Proc. Roy. Soc. London* (1888), **44**, i–xxv.
              F. Darwin (1892).
George Eliot: *Dictionary of National Biography*, **13**, 216–22 (under Cross, Mary Ann).
              Cross (1902).
Galton: *Proc. Roy. Soc. London B* (1911), **84**, x–xvii.
              Pearson (1914–30).
Huxley: *Proc. Roy. Soc. London* (1895–6), **59**, xlvi–lxvi (P) (by Foster).
              Huxley, L. (1900) (P).
Lewes: *Dictionary of National Biography*, **33**, 164–7.
              Smith (1960) (P).
Rolleston: *Proc. Roy. Soc. London* (1881–2), **33**, xxiv–xxvii.
Spencer: *Dictionary of National Biography.* 2nd Supplement, **3**, 360–9.

PART III:
EXPERIMENTAL
PHYSIOLOGISTS 1870–85

9

Experimental physiology

The years about 1870 mark a complete change in British physiology. It is remarkable how little connection can be discerned between the part-time teachers of 1835–70 and the rapid development of British physiology towards its modern form which occurred in the next fifteen years. This was a time of rapid change in the whole system of University education: by means of their own internal outlook and by the outside force of Royal Commission and Act of Parliament, the existing universities were reformed and new colleges and universities formed by which adult education was extended to a wider range of students, including women.

The place of physiology in the forefront of this expansion was ensured by the high qualities of the men whose lives are recounted in this section and who were the immediate founders of modern physiology. Two factors particularly responsible for the growth of physiology were acceptance that physiology must be taught by practical classes, and that physiologists must conduct research on animals.

Teaching by practical classes

The 'Medical Students' Class Guide' published by the *Lancet* in September 1869 shows all London and provincial schools in England as offering a course of 3–6 lectures per week in 'anatomy and physiology', with a parallel course in 'anatomy, descriptive and surgical' and a course of 'anatomical demonstrations'. Substantially this was the same arrangement as in 1830 but there was much more physiology to teach in 1870. Although the courses in 1869 did not separate physiology from anatomy, the requirements of the examining bodies did specify physiology as a separate subject to be passed at the first professional examination; the Recommendations of the General Medical Council included physiology as one of the subjects 'without a knowledge of which no candidate should be allowed to obtain a qualification entitling him to be registered'.

In 1869, many schools offered a course of 1–2 classes per week for half of the year, variously entitled histology, microscopical anatomy, use of the microscope. Only one school, University College, London, offered a course in practical physiology, on Saturday mornings, under Dr M. Foster. Many schools however at that time made no separate provision for the teaching of histology or physiology. In Leeds, when the new Medical School building in Park Street was opened in 1867, it had a Department of Anatomy with museum and dissecting rooms and a Department of Chemistry with a students' laboratory, but no provision for physiology (Anning and Walls, 1982).

By 1881, as set out in the *Lancet*'s 'Medical Students' Class Guide' of 10 September of that year, many London Schools had a course of lectures on physiology, separated from anatomy and also offered a course in experimental or practical physiology. Similar courses were available in Cambridge, Liverpool, Manchester and Sheffield but not in other provincial schools; in Leeds space had become available for physiology when the teaching of chemistry was taken over by the newly formed Yorkshire College, but the teaching of physiology was not yet separated from anatomy. It is apparent from the Leeds prospectus of 1881 that 'practical physiology' in Leeds was entirely histology – the preparation and examination of tissues under the microscope.

The regulations of the examining bodies also moved towards requiring practical work. First to state such a requirement was the Royal College of Surgeons, who after 1870 said that the student must have a certificate of having attended a course of thirty practical classes in general anatomy and physiology, as well as practical classes in chemistry and pharmacy. No similar requirement was formally stated by the University of London or the Society of Apothecaries until after 1885. The status of the examination in physiology increased over the period 1870–85 and by 1880 practical work in physiology, largely histology, had become the rule rather than the exception. To meet this requirement schools had to provide laboratories, equipment and staff.

In other disciplines also, teaching by lectures was becoming supplemented and to some extent replaced by practical classes. At University College, London, practical classes in chemistry already existed by 1852 (see p. 83) and were introduced in physics in 1867 and in applied mathematics and mechanics in 1878 (Bellot, 1929). In biological subjects, Huxley (p. 113) had provided an initiative, which at University College led Lankester (p. 245) and Oliver to introduce practical classes in zoology and botany in 1875 and 1880.

After 1870 the strongest single influence on the growth of physiology in Britain was practical class teaching, introduced at about the same time by Huxley in his classes at the School of Mines (p. 113), by Sharpey, Foster and Burdon Sanderson at University College, London (p. 86) and by Hughes Bennett and his assistants in Edinburgh (p. 97). Perhaps Huxley should be regarded as the central figure in this new method of teaching, because he involved so many young men as his assistants.

Research work

In the biographies of Part II one finds very little research work of importance; indeed most of the medical physiologists of the mid-Victorian period did no original work. They had no laboratory or finance for research work and in any case were fully occupied by their lectures and their clinical work. As clinicians they had no inducement to go and learn methods of research at European centres. The only physiological research in this period was by means of the microscope − such as Bowman's description of the renal glomerulus, Waller's description of Wallerian Degeneration and the work of Wharton Jones and Lister on inflammation.

As yet there was in Britain virtually no research on animals other than frogs. Anaesthesia was still little known and in 1870 was itself the subject of study by Richardson and others. The older physiologists, for example Sharpey, still felt a dislike of 'vivisection' which was left over from the days of Magendie (p. 23); the younger physiologists, such as Burdon Sanderson, having observed the more mature work of Claude Bernard and Ludwig, saw the need for experiments on animals and after 1870 these were made acceptable by the use of anaesthesia. Thus Schäfer, Gaskell, Langley, in 1875 the new generation of physiologists, were the first to use the anaesthetised animal as a main experimental preparation.

These men were also the first full-time physiologists for whom experimental work was part of their duties. They demanded laboratories, equipment and staff for research as well as teaching.

Physiology 1871

A convenient illustration of the state of teaching in physiology in 1871 is provided by twelve lecture-demonstrations given by Rutherford at King's College, London and published in the *Lancet* 1871−2. Rutherford (p. 188) was then Professor of Physiology at King's, with Ferrier (p. 190) as lecturer. For the session 1871−2, the classes at King's were lectures on anatomy and physiology by Rutherford on Monday, Wednesday, Thursday and Friday at 4 pm, fee £7 7s and practical physiology by Ferrier on Tuesday and Saturday at 11.15, fee £5 5s. There were also lectures daily at 9 am on descriptive and surgical anatomy and daily anatomical demonstrations.

Rutherford summarised the teaching of practical physiology at King's in the following note printed after one of his lectures (*Lancet*, 1871, i, 707):

As some readers may be puzzled to know why these lectures are headed Practical Physiology, it may be well to explain that I have for some years been teaching Practical Physiology on the following plan:- The class is divided into three sections − I. Practical Histology; II. Practical, Physiological, and Pathological Chemistry; and III. Experimental Physiology. In section I, every student works with his own microscope. He demonstrates, draws, describes, and mounts the various tissues of the body. In section II. he performs the qualitative analysis of most of the fluids and solids of the body, together with the quantitative analysis of the urine. In section III. he merely assists me in performing experiments such as those detailed in these lectures. Section III., therefore, chiefly consists of demonstrations made by the teacher with the aid of the students. The teaching in this section is, of course, not *practical* in the truest sense of the term. Some are inclined to advise that in experimental physiology every student should perform all the experiments. I have to some extent tried this, but was compelled to abandon it. It is impracticable, and for various reasons unadvisable. As far as I am aware it is not a mode of teaching experimental physiology adopted in any school in Europe. I think we do enough if we show students how to perform physiological experiments, and afterwards superintend and guide the attempts of the few who may desire to undertake any research with a view to discover new facts. – W. R.

At the end of the first lecture it was stated that the demonstrations are confined to the frog and the rabbit and that the animals are previously narcotised by opium or chloral.

This account of practical physiology by Rutherford was like courses at University College and Cambridge, except that Rutherford was exercising his pleasure in elaborate demonstrations. The course at University College is set out in *Handbook for the physiological laboratory* by E. Klein, J. Burdon Sanderson, M. Foster and T. Lauder Brunton (1873) and that at Cambridge in a book for the use of students, *Practical physiology* by M. Foster and J. N. Langley (1876). In each place the course comprised histology, chemistry and experiments, although the sequence varied. These were the first of many such student handbooks.

The twelve lectures by Rutherford which were published in the *Lancet* included seven on nerve and muscle in which contractions of the isolated gastrocnemius muscle were recorded by lever and kymograph and the sciatic nerve was stimulated by an induction coil. These experiments persisted in use in student classes until they were replaced about 1960 by electronic stimulators and cathode ray oscilloscopes. Five of the lectures dealt with the circulation; a model of the circulation was demonstrated and the effects were shown of the vagus on the heart and of the sympathetic on the blood vessels of the ear of the rabbit. Arterial blood pressure was recorded with a mercury manometer. Instruments such as Ludwig's stromuhr were shown to the class. The University College and Cambridge books covered similar experiments. In these practical classes and in Foster's textbook of 1877 (p. 31), there was something close to material still taught as the basis of modern physiology.

Physiological laboratories 1871

Practical classes in Physiology imply that there was equipment available, such as recording levers, mercury manometer and kymograph. The classes in 1870 were far removed from Sharpey's early days at University College, when in his lectures he would hold up and rotate his black stove-pipe hat to explain the working of the kymograph.

Practical classes also involve staff. After 1857 Sharpey had first Harley, then Foster as Lecturer in Practical Physiology and after 1870, Burdon Sanderson was Professor of Practical Physiology and Schäfer was the first Sharpey Scholar and Assistant to the Professor. Physiology had then achieved in University College the status of a separate subject with its own staff, as compared to the time of Sharpey's appointment as Professor of Anatomy and Physiology; then physiology was regarded as a minor part of lectures in anatomy given by surgeons.

Laboratories were an absolute necessity for the new type of physiologist and regular items on the agenda of the governing bodies at University College, King's College and Cambridge were the requests of the Professors of Physiology for laboratory space. Research work was beginning at these places and there had to be laboratories, equipment and staff for this purpose also.

A model of the ideal Institute of Physiology was available. H. P. Bowditch (1840–1911), a graduate of Harvard University and soon to become its first Professor of Physiology, had been working for two years with Ludwig. *Nature* (1870, *3*, 142) and the *Lancet* (1871, i, 65) published his account of the new, purpose-built laboratory which had just been opened in Leipzig for Ludwig and recommended this as the ideal towards which developing physiology departments should aim.

The new building in Leipzig consisted of a central block and two wings, each main face being about 120 feet long. The right wing was the microscopical department, the left wing the chemical department and the centre block was devoted to experimental physiology. For histology there was a general room for the teaching of histology, a room for advanced students and a private room for the assistant in microscopy. There were two main chemical laboratories with smaller rooms for special processes. There were three laboratories equipped for experimental physiology, furnished with operating tables, bellows for artificial respiration, registering apparatus for the pressure of the blood, water baths with controlled temperature, apparatus for injection and glass cases for storage of apparatus. A separate room was set aside for experiments using quicksilver and this had apparatus for measuring activity of respiration in man and animals. In addition to the permanent apparatus, new apparatus was continually ordered for special investigations. There was also a skilful mechanic living in the laboratory.

There were ancillary rooms in the basement, containing a small gas engine to drive the equipment, workshop, store rooms and rooms in which small animals were kept. In the courtyard a separate building contained necessary arrangements for experiments on horses and other larger animals, an aviary and a small fishpond.

Extending from the central block into the space between the two wings was a lecture theatre capable of accommodating 100 students. Tables ran on a small rail road in front of the seats enabling the lecturer to demonstrate his experiments very conveniently.

Above the laboratories, a second story contained the rooms of Professor Ludwig and his family and those of other persons connected with the laboratory.

In Bowditch's time, Ludwig had as his assistants, Professor Schweigger-Seidel in microscopy, Dr Hüfner in chemistry, and Dr J. J. Müller in physics. They gave lectures on their own subjects and Ludwig himself lectured five times a week in physiology. There were nine research students. Ludwig devoted his whole time to supervising the work of these students, making no independent investigations himself. Results were usually published solely in the name of the student, only sometimes with that of Ludwig added. Mostly publication was in a pamphlet, of about 200 pages each year.

The funds for the new laboratory were provided by the liberality of the Government of Saxony, which bore *all* of the expenses of running the laboratory, so that the institution was absolutely free of charge to the student. Professor Ludwig could welcome any student of any nationality who was desirous and capable of original investigation.

The publication by the *Lancet* and *Nature* of Bowditch's account of the new laboratory was clearly intended to set this up as a model for future developments in England.

Bowditch's article in *Nature* of 22 December 1870 began by saying that there were then no physiological laboratories in Great Britain. In the next number of 5 January 1871, Burdon Sanderson qualified this statement by listing the laboratories he knew to be then active. Firstly Sharpey for many years had had a laboratory at University College, London, now consisting of three rooms, one for students, one for research and the preparation of demonstrations, one for purposes requiring a separate apartment. In addition to work done by Sharpey and staff of University College, this laboratory was open, on payment of fees, to men outside University College. Burdon Sanderson cites work recently done by a Committee of the Medical and Chirurgical Society on apnoea and subcutaneous injections, work on the effect of respiration on the heart and his own work on the transmission of cholera. Secondly, there was a laboratory at King's College, London which Professor Rutherford was to develop further. Thirdly, there were laboratories in Edinburgh; the University Physiological Laboratory of Professor Bennett; laboratories of the Professors of Materia Medica (Christison) and Medical Jurisprudence (Maclagan) and of Gamgee, Lecturer in Physiology at the Royal

College of Surgeons. Burdon Sanderson mentions Rutherford's work in Edinburgh on the action of the vagus of the heart, and Brunton's on digitalis, amongst other work carried out in Edinburgh in the previous few years.

As yet there was no laboratory in Cambridge; one was soon established by Foster.

Some insight is given into the situation of research in 1870 when Burdon Sanderson stated that, although these laboratories were open to anyone who wished to work in them, there was quite a high charge for an individual to have access to them. In contrast, Ludwig's laboratory was open without charge to anyone whom Ludwig thought capable of using it profitably.

When, on 5 January 1871, Burdon Sanderson listed physiological laboratories in Great Britain, he did not mention the Brown Institute, which was opened in December 1871 with Burdon Sanderson himself as Professor-Superintendent.

Brown Animal Sanatory Institute

In the will of Thomas Brown of London and Dublin who died in 1852, there was a bequest to the University of London for the treatment and study of diseases of quadrupeds and birds useful to man. However the will was interpreted to mean that the money could not be used to build an institution, but only for treatment and research. The will was further complicated by a proviso that if, after nineteen years, the money was not being used, it would pass to Trinity College, Dublin for the establishment of Professorships in Oriental Languages. To solve the dilemma, the University of London promoted a bill to allow the setting up of studentships and scholarships in Veterinary Medicine; this passed the House of Commons but was defeated in the House of Lords, perhaps by a turn out of Irish peers. Eventually, just in time, a City merchant was induced by Burdon Sanderson to provide £2,700 to buy land and equip the institute so that the bequest could be used for its primary purpose. The trust had then accumulated to £32,334 15*s* in three per cents.

The Institute is described in the *Lancet* of 4 November 1871 as ready for occupation within a month. It was situated in Wandsworth Road, close to the Nine Elms Railway Station, on about 1½ acres of land. The buildings were in three blocks. The first was for animals under treatment and observation and consisted of a stables with five stalls, five rooms for other animals. a post mortem room and storage for food, etc.; a semi-detached group of dog kennels was connected with this building. The second block was devoted to scientific purposes as a pathological laboratory, containing five work rooms and ancillary services. The third block was two dwelling houses. The Institute thus provided animal research facilities larger than existed elsewhere in London and particularly it was the only institute allowing research on large animals.

The staff consisted of the Professor-Superintendent (Burdon Sanderson, p. 141), an Assistant Professor (Klein, p. 155), a veterinary surgeon and three servants. A duty of the Professor-Superintendent was to give each year a course of five popular lectures. The general management was by a committee of the Senate of the University of London, with Sharpey as the first Chairman.

The Institute was to serve two purposes. Firstly, for a fee, it would treat sick animals brought to it. Secondly, it was to investigate problems in animal pathology, but at first physiology was rather excluded as not being in the terms of the bequest. It would take advanced students who on payment of their fees, could learn advanced pathology and carry out original experiments. The conditions were laid out in the *Lancet* (1872, i, 514). On the recommendation of the Professor-Superintendent, admission to the laboratory would be granted to persons who had appropriate scientific training with the conditions that their work must be only on a subject approved by the Superintendent; results were to be published only with his approval and with acknowledgement of the Brown Institute. Each student was to provide himself with those instruments which are required for his exclusive use, defray the expense of materials and contribute the sum of ten shillings per month, payable in advance, towards the incidental expenses of the laboratory.

Lancet (1875, i, 740) contains Burdon Sanderson's report for 1874. In fulfilling its function of treatment of quadrupeds useful to man, it had treated 2971 horses, 922 dogs, and 331 other species. There is comment that most of the sickness in horses was due to injury and ill treatment. A special note is made of 63 dogs with rabies. Scientific research had been undertaken at the expense of the Government at the instance of the medical officer of the Privy Council. This was into diseases destructive of stock in this country. Another subject brought to notice was grouse disease.

The superintendents in succession were: Burdon Sanderson (1871–8), Greenfield (1878–81), Roy (1881–4), Horsley (1884–90), Sherrington (1891–5), Rose Bradford (1896–1903), Brodie (1903–8) and J. F. Twort (1909–44). Twort was still superintendent when the Institute was destroyed by bombing in 1944. Although the Institute had the main function of investigating disease, only Greenfield and Roy became pathologists and Twort was a bacteriologist; the others became eminent physiologists and clinicians.

Following the destruction of the buildings, after much discussion and legal advice, the University of London decided to use the accumulated assets to support a Fellowship in Veterinary Pathology at the Royal Veterinary College. The legal arguments took time and it was not until 1971 that half of the £96,000 then available was used to found the Brown Fellowship and half, by out of court agreement, was used to support lectureships in Eastern Languages at Trinity College, Dublin.

A full account of the Brown Institute, its foundation, directors, scientific work and demise has been given by Wilson (1979).

References

Anning, S. T. and Walls, W. K. J. (1982). *A History of the Leeds School of Medicine: One and a Half Centuries, 1831–1981*. Leeds University Press.
Bellot, H. (1929). *University College, London 1826–1926*, University of London Press.
Wilson G. (1979). 'The Brown Sanatory Institution', *Journal of Hygiene*, **82**, 155–76, 337–52, 501–21 and **83**, 171–97.

10

Physiologists at University College, London

The developing Department of Physiology

In 1871, laboratory accommodation for physiology at University College consisted of three rooms (see p. 86) and this was increased in 1874 by opening the Jodrell Laboratory in the old Materia Medica room. Laboratory space was greatly increased when the north wing was built in 1880 with the top floor allocated to physiology. For some purposes laboratory space was also available to Burdon Sanderson in the Brown Institute (p. 136). Even so, to have a laboratory for his own use, Burdon Sanderson established a laboratory in his own house in Howland St, off Gordon Square (p. 143).

A first demand on the laboratory facilities of University College and its staff was the teaching of undergraduate students. Practical physiology was now required by the College of Surgeons, and the number of students in the practical classes therefore increased. In 1871 the teaching staff in physiology was two professors, Sharpey and Burdon Sanderson, and the first Sharpey Scholar, Schäfer. Sharpey still gave a course of lectures in General Anatomy and Physiology, although it seems Burdon Sanderson did some of these lectures. Burdon Sanderson and Schäfer were responsible for the practical classes, with an assistant, F. J. M. Page, who taught physical chemistry. When Sharpey retired in 1874, Burdon Sanderson became Jodrell Professor, and Schäfer became Assistant Professor but the teaching in the practical classes remained unchanged, Page teaching chemical physiology, Schäfer histology and Burdon Sanderson experimental physiology. The lectures were now less important than in Sharpey's earlier times. In addition the teaching was assisted by a succession of Sharpey Scholars – G. A. George appointed in 1873, W. Murrell in 1875, P. Geddes in 1877, W. North in 1879 and F. Gotch in 1881.

Research was done both by the staff of the department and by other workers. Many of these were members of the clinical staff of University College Hospital, who had been introduced to physiology in their student days and were allowed space in the laboratory for their own original work. The most notable of these was Sydney Ringer, who was a physician in the hospital and Professor of Materia Medica and Therapeutics and later Professor of Medicine. In the first four volumes of the Journal of Physiology (1878–1882) were fifteen papers by Ringer, including papers on the action of salts on the frog's heart, from which Ringer's solution evolved (p. 155). In some of these and later papers, Ringer had as collaborators E.A. Moreshead, H. Sainsbury or D.W. Buxton who were assistants to the Professor of Materia Medica or Medicine. G. Harley (p. 151) was another physician of University College Hospital who sometimes worked in the Physiological Laboratory.

There were also men whose undergraduate education had been elsewhere but came to work at University College drawn by the reputation of Burdon Sanderson. Such was William Osler (1849–1919) who in 1872, then aged twenty-three, arrived in London from Montreal seeking further medical education. He had some idea of specialising in ophthalmology and consulted Bowman, who recommended that in any case he should work for about a year at University College with Burdon Sanderson. Thus in 1873 he was engaged in the microscopic examination of blood, producing firstly a paper published in the Quarterly Journal of Microscopic Science, 'On the action of certain reagents, atropia, physostigma and curare on the colourless blood-corpusles' and a second paper communicated to the Royal society by Burdon Sanderson, 'An account of certain organisms occurring in the liquor sanguinis'; this was work done with Schäfer and describes blood platelets. Osler's experience in University College and contact with Klein (p. 155) made him an accomplished microscopist and he was well qualified to be lecturer, and in 1875 professor, at the Institutes of Medicine at McGill University, Montreal. Some thirty-three years later Osler succeeded Burdon Sanderson as Regius Professor of Medicine at Oxford.

Between 1874 and 1882 papers were published from University College by about twenty authors, most of them visitors. The list includes Romanes, Horsley, Mott, Gotch, Rose Bradford, Bayliss, A.D. Waller, men who in the years after 1885 continued to make scientific contributions relevant to physiology, although professionally engaged in other fields. Some lectured for a period in physiology before devoting their full time to clinical work.

The research topics ranged between the work of Burdon Sanderson on carnivorous plants, Schäfer and Romanes on the swimming bell of Medusa, Schäfer on insect muscle and at the other extreme, work by Page on CO_2 production by conscious dogs; this was actually done at the Brown Institute. Much of the early work was microscopic investigation, for example, the work of Schäfer and of Osler referred to above. Experimental work on heart and muscle by Burdon Sanderson, Ringer and others was on frogs. There were a few experiments on cats or rabbits anaesthetised with morphia and ether or chloroform.

Under Burdon Sanderson, the Physiology Department had by 1880 acquired the characteristics of a modern department. There was the basal commitment of teaching medical students and also some science students. The staff undertook their own research, encouraged their own students, and there were also research students and workers from outside.

The following biographies recount the lives of the two professors who were founders of the Department of Physiology at University College, London and of men who were established research workers in the College before 1885. Most of these men were active in physiology for many years after 1885 and the later parts of their lives are also given here.

Professors of Physiology

Sir John Scott Burdon Sanderson, FRS (1828–1905)

John Burdon Sanderson was born at North Jesmond, Northumberland, into a family of considerable wealth and with men of distinction on both his mother's and father's sides. He was educated privately in their home at West Jesmond in the Northumberland of swelling moors and rocky streams. He always had a great love of the wild countryside and acquired a wide knowledge of botany.

By family example he could well have studied law, but he chose medicine and in 1847 went to Edinburgh and qualified as Doctor of Medicine in 1851, being particularly influenced by Hughes Bennett, Professor of the Institutes of Medicine. During his student period Burdon Sanderson read papers on Vegetable Irritability and on the Metamorphosis of Red Blood Cells, both reporting some experimental observations.

After taking his degree, Burdon Sanderson went to Paris particularly to study organic chemistry and there worked in the laboratory of Wurtz, studying substances like kreatine and uric acid. Associates in Edinburgh and Paris were G. Harley (p. 151) and W. Pavy (p. 209). In Paris they all went to experimental demonstrations given by Claude Bernard for whom Burdon Sanderson subsequently had so much reverence that he always kept a bust of Bernard on the shelf above his desk. They also met Magendie in March 1852. Burdon Sanderson's notebooks contain much detail about experiments with Bernard, such as 'performed operation for obtaining pancreatic juice on a rabbit: did not succeed in introducing the cannula'.

From Paris he went to London to set up as a physician, possibly because he married a Londoner, Ghetal Herschell, whose brother Farrer subsequently became Lord Chancellor. Burdon Sanderson in 1854 was appointed Medical Registrar at St Mary's Hospital, Lecturer in Botany and later Medical Jurisprudence. In 1854 he contributed to Todd's Cyclopaedia (p. 36) on 'Vegetable Reproduction'. In 1856, he was appointed Medical Officer for Health for the district of Paddington, thus beginning the first of four major phases of his career, that of pathology and public health, particularly in relation to infectious diseases.

Burdon Sanderson was Medical Officer for Paddington for eleven years and served the district in the same way as Simon (p. 41) had served the City of London a few years earlier. He visited slaughter-houses, cowhouses and bakehouses, looked at nuisances and reported on them; he forced improvements in the condition of Regent's Canal and was Medical Officer for Health through two cholera outbreaks. Burdon Sanderson and Simon had much in common in their scientific background and outlook and in their practical response to immediate sanitary problems. Simon was then Medical Officer of the General Board of Health and he appointed Burdon Sanderson as a medical inspector under the Privy Council, involving him in special investigations. Thus between 1857 and 1860 Burdon Sanderson investigated epidemics of diphtheria in seventy centres in England, inspected vaccinations, investigated outbreaks of smallpox and cholera, and reported on cerebrospinal meningitis in Germany.

At that time, cholera, tuberculosis, smallpox, typhus, and typhoid were suspected of being infectious but there was as yet no bacteriological study. The Privy Council, advised by Simon, set up committees (of which Burdon Sanderson was a member) to investigate some of these diseases, particularly to demonstrate the manner of their transmission. Burdon Sanderson found that the infective agent of cattle plague did not diffuse through parchment paper and could not be seen under the microscope. Some of this work was carried out in Sharpey's laboratory at University College. Burdon Sanderson and Simon in 1870 were the leading pathologists studying infectious diseases and were also very practical men in matters of public health. Recognition of Burdon Sanderson's status as a pathologist was his selection to conduct a post-mortem examination on Napoleon III who died at Chislehurst in 1873.

Burdon Sanderson's writings – reports in official documents on epidemics and sanitary conditions and reports of special investigations – were widely acclaimed. Huxley in 1870 noted an appendix to one of Simon's official papers 'On the intimate pathology of contagion' by Burdon Sanderson and called it 'one of the clearest, most comprehensive and well reasoned discussions which has come under my notice'.

While Burdon Sanderson was Medical Officer of Health he was also a practising physician, although without time for many private patients. He was physician to the Brompton Hospital for Consumption (1859–1871) and Assistant Physician at the Middlesex Hospital (1863–1870). At the Middlesex he lectured on physiology and pathology for four years (1866–70).

In 1870 he gave up his clinical and public health work to devote himself to full-time scientific work in pathology and physiology, thus beginning the second phase of his career. The opportunity came by his appointment as Professor of Practical Physiology and Histology at University College, to succeed Michael Foster who had gone to Cambridge (p. 168). Burdon Sanderson was already working in Sharpey's laboratory on some of the special investigations commissioned by the Privy Council and other bodies; he had been elected FRS in 1867. When Burdon Sanderson was appointed as a second professor in Sharpey's laboratory he was already a man of standing and was the natural choice for the post of Professor-Superintendent of the Brown Institute (p. 136). He was then in effective control of about three-quarters of the physiological laboratory space available in London. In addition he fitted a laboratory in his own home. Gotch, in speaking of his first close contact with Burdon Sanderson, said:

I called on him in his house in Gordon Square. Our interview took place in the dining-room, and when he realized that I thought of trying to do a little physiological work of an advanced kind, his response was prompt and singularly to the point. It consisted in leading me into a small back room full of apparatus, where a gentleman unknown to me was seated in what seemed to be an impenetrable jungle of wires.

The gentleman was F. J. Page, working on 'The time relations of the excitatory process in the ventricle of the heart of the frog', a paper published jointly by Page and Burdon Sanderson in the *Journal of Physiology* in 1880. Gotch, then a student, later became Burdon Sanderson's assistant at University College and Oxford and succeeded him as Professor of Physiology at Oxford.

At University College, at first as an outside worker, then as assistant and after 1874 as Jodrell Professor, Burdon Sanderson was an active experimenter. In 1862–84 he was a leader in an investigation for the Royal Medico-Chirurgical Society on the best means of artificial ventilation. He designed a stethograph for measuring movements of the chest wall by which the committee could conclude that the Sylvester method was more effective than that of Marshall Hall (p. 17). This instrument also led to a Croonian Lecture (1867) to the Royal Society on 'The influence exercised by the movements of respiration on the circulation of the blood'. In 1873, Burdon Sanderson, Foster, Brunton and Klein published *The Handbook of the Physiological Laboratory*, at that time the most detailed and extensive description of physiological experimental methods; Burdon Sanderson was himself responsible for the section on circulation and respiration. About 1870 he began investigations of the electro-physiology of the frog heart and papers with Page in 1880 and 1883 accurately describe the spread of electrical negativity. In 1873, on the suggestion of Charles Darwin, Burdon Sanderson began to study the leaf movements of insectivorous plants, including the associated electrical changes induced by movements of the sensitive hairs; the result was three papers on the electromotive properties of the leaf of *Dionoea muscipula* published in 1877, 1882 and 1888.

While Jodrell Professor, Burdon Sanderson was in 1876 one of the originators of the Physiological Society, the initial meeting being held in his house (p. 262). He was a regular attender at meetings and became regarded as its most senior and distinguished member.

This second period of Burdon Sanderson's career ended in 1883, when Rolleston, Linacre Professor of Anatomy and Physiology in Oxford, died and the Waynflete Chair in Physiology was instituted. Burdon Sanderson's reputation was very high and some members of the University saw his appointment as the best way of introducing the new medical science at Oxford. Rolleston (p. 109) was an old-style teacher, giving excellent lectures on comparative anatomy but without appreciation of experimental work. Many in Oxford did not want a physiologist, the Waynflete Chair having been imposed on the University by the report of the University Commission of 1877. There was active opposition which came to a head when Convocation was confronted with a proposal to spend £10,000 on the erection of a Physiological Laboratory. Opposition to this proposal became an antivivisectionist campaign against Burdon Sanderson and the decree was carried by the narrowest of majorities, eighty-five to eighty-two. Other decrees were similarly opposed, with the childish last course of opposing the grant for gas, water, coal and running expenses of the laboratory established by earlier decrees. This storm went on for about a year and afterwards those who remembered it said that

Oxford and Physiology should be thankful for the quiet firmness, dignity and self-control shown by Burdon Sanderson (Church, *BMJ*, 1905, ii, 1487).

Thus a laboratory was established at Oxford in which Burdon Sanderson continued his work on the electrophysiology of muscle, using an improved capillary electrometer. He also studied the electrical organs of fishes, taking his apparatus to Arcachon in 1886, St Andrews in 1887 and Plymouth in 1888; with Gotch two papers on the Electrical Organ of the Skate were published in the Journal of Physiology in 1887–8. His work on electrophysiology was collected into the Croonian Lecture to the Royal Society in 1889: 'The relation of motion in Animals and Plants to the electrical phenomena associated with it'.

As colleagues in Oxford, Burdon Sanderson had firstly Gotch, who came with him from University College and returned to succeed him as the Waynflete Professor. Others were Dixey, Buckmaster, Kent, Pembrey, Lorrain Smith and Leonard Hill. In 1886 he was joined by his nephew, J. S. Haldane, who taught Chemical Physiology and was to form in Oxford a major school of respiratory physiology.

Burdon Sanderson had also the general object of establishing Oxford as a school of scientific medicine and in this he had the support of Sir Henry Acland, Regius Professor of Medicine. Burdon Sanderson was largely responsible for the appointment firstly of a lecturer (1885) and then a Professor of Human Anatomy, with the necessary laboratory. Similarly he was responsible for the formation of a lectureship in pharmacology. Later as Regius Professor he worked to establish pathology as a subject in the University.

It was the natural culmination of his conception of Oxford as a school of scientific medicine that when Acland resigned in 1895, Burdon Sanderson reluctantly left his work in physiology for the more general work of Regius Professor of Medicine. In this fourth phase of his career, his first effort was the promotion of pathology, with James Ritchie being appointed lecturer, then reader, and finally the first professor, with his own laboratory. Burdon Sanderson also helped the development of the Radcliffe Infirmary as a teaching school. In 1904, when he resigned because of ill-health, there was an established school of medicine with chairs and lectureships in physiology, human anatomy, pharmacology, pathology occuped by men of his choice. A final satisfaction was the appointment of William Osler, a pupil of his at University College (p. 140), to the

Regius Professorship. Burdon Sanderson was knighted in 1899 at the same time as Michael Foster (p. 170).

John Burdon Sanderson died in Oxford in November 1905 and was buried at Wolvercote Cemetery, three miles from Oxford, after a service in the chapel of Magdalen College of which he was a fellow.

Amongst obituary notices, the most striking is that in the British Medical Journal, 1905 (ii). It is written by seven chosen contributors: F. Gotch, Professor of Physiology, Oxford; E. Klein, pathologist; J. Ritchie, Professor of Pathology, Oxford; Edward Seaton, Medical Officer to the Surrey County Council; Sir William Church, President of the Royal College of Physicians; Arthur Thompson, Professor of Human Anatomy, Oxford; William Collier, British Medical Association. Added were personal tributes from friends – T. Clifford Allbutt, Regius Professor of Medicine, Cambridge (an old friend), and physiologists J. G. McKendrick (p. 202), Sir Victor Horsley, A. D. Waller (p. 215), W. D. Halliburton, Leonard Hill, M. S. Pembrey, C. S. Sherrington and W. Stirling. These were of the next generation of physiologists and their tributes truly represent his great contribution as a physiologist, in organization, in actual experimental work and in the training and inspiration of the next generation.

Such personal tributes all give the same picture of the man. In the words of Leonard Hill:

The striking thing about Burdon Sanderson was his personality. He was a big man and looked at things in a big way. He had an enthusiasm for all knowledge, an interest in all the higher pursuits of man. This enthusiasm he had the power to communicate to students who in any measure possessed the same sympathetic and large nature. He was not as a teacher fitted to the task of making his subject clear and simple to the ordinary medical student, but was rather a source of inspiration to the highest class of student. ... Of his absent mind, many are the stories, and some apocryphal; but I myself have seen him wiping the board with his handkerchief and smoking a big stick of chalk. The latter he pulled out of his mouth and crossly regarded because it refused to draw ... At breakfast one day when a lady passed back her plate and requested the Professor to help her less liberally to bacon, he piled some fish on top of the bacon, and returned the plate with that sympathetic smile which charmed all who knew him.

Students regarded him as one who could sympathetically understand their problems, take part in their activities and lead them towards his own high scientific standards. Part of the fun of his lectures was not knowing what 'The Burder' might do next.

Sir Edward Albert Sharpey-Schafer, FRS (1850–1935)
Born Edward Albert Schäfer he was the son of J.W. Schäfer of
Highgate and Hamburg, and grandson of a famous musician of
Hamburg. Edward Schäfer was educated at Windsor and entered the
medical school of University College in 1868. Here he came in contact
with Sharpey and with Foster, and was awarded scholarships in
zoology and in anatomy and physiology. His merit was recognised
by Sharpey and he was elected in 1871 as the first Sharpey Scholar.
This post carried teaching duties, so Schäfer became the assistant
to Burdon Sanderson, recently appointed Professor of Practical
Physiology under Sharpey. He continued medical studies to become
MRCS in 1874. He went also to Leipzig to Ludwig's Institute and
acquired great respect for Ludwig, 'whose very appearance Schäfer
seemed to acquire'.

When Sharpey retired and Burdon Sanderson became Jodrell
Professor, Schäfer in 1874 became Assistant Professor. Schäfer
succeeded Burdon Sanderson as Jodrell Professor in 1883 and so
remained at University College until 1899, when he went to Edinburgh
on the death of Professor Rutherford. Schäfer retired in 1933, having
served Physiology at University College for twenty-eight years and
in Edinburgh for thirty-four years. He died in 1935.

In his early days at University College, Schäfer was pupil and
assistant to Sharpey, Foster and Burdon Sanderson. When Foster
went to Cambridge, he asked Schäfer to accompany him as his
assistant but Schäfer chose to remain at University College. Later,
Schäfer said that it was Foster who persuaded him to devote his life
to physiology, but Schäfer gave his greatest veneration to Sharpey.
Sharpey had no family and Schäfer sought to perpetuate the name
by calling his son Sharpey-Schafer. However the son was killed in
action in a Q-boat in 1918 and Schäfer himself took the name of
Sharpey-Schafer 'to emphasise his indebtedness to one who inspired
his early work, a great scientist and staunch friend'.

When Schäfer entered University College, Sharpey and Burdon
Sanderson had established practical class teaching and had established
experimentation as the method of physiological research. In retrospect
one can see that what was required of Schäfer was that he should
consolidate the foundations laid by the older men, by himself doing
real experimental work and by teaching the experimental method to
students. The contribution of Schäfer is, therefore, in his own
experimental work, and in his students.

At first his contribution was in histology. He taught practical histology. He wrote the sections on microscopical anatomy for *Quain's Anatomy* (p. 28) of which Thane was then the editor of the anatomical sections. In 1885 he published his own book, *Essentials of Histology*, a book used by medical students through many editions; Schäfer himself wrote it until the twelfth edition in 1929: thereafter it was edited by Carleton until the fifteenth edition in 1949, a total of 123,000 copies. Schäfer's first research work in the 1870s was histological, on the wing muscles of insects and on the histology of fat absorption in the villi of the small intestine. In 1878, at the young age of twenty-eight, he was elected FRS.

Papers in the *Journal of Physiology* in 1881 deal with proteid substances of the blood and the alkalinity of the blood. In about 1880 Schäfer began researches on cerebral localisation of function, this including papers published in 1886 with Horsley on the character of muscular contractions produced by stimulation of various parts of the motor tracts. The importance of this work was attested by Sherrington in the first Sharpey-Schafer Memorial lecture in Edinburgh in 1935.

In 1894, Schäfer and Oliver demonstrated to a meeting of the Physiological Society in Edinburgh the large rise in blood pressure produced by intravenous injection of extract of the suprarenal gland. Reports on the effects of extracts of other glands soon followed and so Schäfer became a great developer of endocrinology; in fact Schäfer gave many of the names to this subject including the words *endocrine* and *autocoid* and was largely responsible for the word *Insulin* describing the then unknown anti-diabetic substance from the Islets of Langerhans. He gave a series of Lane Lectures in Stanford University, published in 1916 as a book, *The Endocrine Organs*, which for many years was a standard source book, with a further edition in 1924–26.

Schäfer also worked on the blood capillaries of the liver, the action of chloroform, the effects of section of the vagi and sympathetic nerves, the pulmonary circulation, and the action of intercostal muscles. In 1903 he described Schäfer's method of artificial respiration and compared it with other methods and so made his contribution in succession to earlier physiologists such as Marshall Hall (p. 17) and Harley, Burdon Sanderson and Horsley (p. 143).

Schäfer's standing as an experimental physiologist became such as to allow him to edit a *Textbook of Physiology* intended for advanced students and teachers. The first volume was published in 1898 when he was still Jodrell Professor at University College, the second volume in 1900 after he had gone to Edinburgh. Schäfer himself wrote chapters on 'The Blood', 'The Nerve Cells', 'The Cerebral Cortex' and other chapters were written by J.S. Edkins, A. Gamgee, W.H. Gaskell, F. Gotch, A.A. Gray, W.D. Halliburton, J.B. Haycraft, Leonard Hill, F.G. Hopkins, J.N. Langley, J.G. M'Kendrick, B. Moore, D. Noël Paton, M.S. Pembrey, E. Waymouth Reid, W.H.R. Rivers, J. Burdon Sanderson, C.S. Sherrington, and E.H. Starling, each dealing with topics to which he had given special attention. This was the first advanced textbook in English, written by specialists, nearly all of whom were 'new' physiologists of the era after 1885. In his preface, Schäfer refers to Herman's *Handbuch der Physiologie*, a German book published twenty years earlier, as the only parallel.

Because of the mass of research work requiring publication, Schäfer founded in 1908, with the help of others, the *Quarterly Journal of Experimental Physiology*; the story of the foundation of *QJEP* is recounted by Whitteridge (1983) to mark the seventy-fifth anniversary. Schäfer was chief editor with Gotch (Oxford), Halliburton (London), Sherrington (Liverpool), Starling (London) and Waller (London) as co-editors; in fact Schäfer did most of the editing himself. As editor he allowed contributors unusual freedom of exercise of their own judgement — 'Their readers are their judges, not I.' This was in deliberate contrast with Langley and his *Journal of Physiology* (p. 177) and is a dictum which modern editors might well take to heart. When he resigned his chair in 1933, he retired from the editorship which then passed to a committee headed by I. de Burgh Daly, his successor in the chair, and Daly was followed by succeeding professors in Edinburgh, until in 1980 *QJEP* was taken over by the Physiological Society.

Sharpey-Schafer was the first man who pursued physiology throughout a working life. From 1871, when he was appointed Sharpey Scholar, until his retirement in 1933, a total of sixty-two years, he was a physiologist. He taught the undergraduates, carried on his research and encouraged the efforts of research workers. As a professor he efficiently organised his department so that personal obtrusiveness was unnecessary, yet he was always available to staff and students. His logical and inductive mind proved of the greatest assistance in the deliberations of the governing bodies in London and in Edinburgh. Indeed he can be cited as the first, and an ideal, modern professor.

His scientific achievements were recognised by many invited lecturerships and by honorary degrees of many universities. The few official positions he held were directly connected with his scientific work. From 1895 to 1900 he was general secretary and in 1912 President of the British Association. He was knighted in 1913. In 1923, he was President of the International Congress of Physiology held in Edinburgh. But his great devotion was to the Physiological Society. As a young physiologist in 1876 he was a founder member (p. 262), and throughout his working life most of his results were first reported at meetings of the Society. At its fiftieth anniversary in 1926 he was the only surviving member of the nineteen men who met to form the society at Burdon Sanderson's house in 1876. He then wrote *History of the Physiological Society during its first 50 years, 1876–1926*, based on the minutes and his own memories. Much of what we know of physiology before 1900 comes from this book. The Society presented him with a leather-bound copy of the book and elected him an honorary member.

As has been said above, Sharpey-Schafer's true memorial lies in his published papers, the work of his students and their spread to all countries. A tribute was the publication in 1933 of a number of the *Quarterly Journal of Experimental Physiology* dedicated to him and consisting of new work by twenty-nine past and present members of his staff in Edinburgh. This was in recognition of his eightieth birthday and his foundation and service to the *Quarterly Journal*. In fact he did not live much longer, and it then became his most eloquent obituary.

His biographers tell no anecdotes of him. They tell of a man, often apparently stern and somewhat awe-inspiring, intolerant of intellectual dishonesty but a real and willing friend to the honest student. At home he was a genial host, a good talker full of illustrative anecdotes from his wide experience. He was a keen and good golfer on the course near his home at North Berwick. W. A. Bain, later Professor of Pharmacology at Leeds, was his last colleague, and Bain wrote in an obituary note:

To most Sharpey-Schafer is known only by his writings; to some thousands he was known as a teacher; to some hundreds as a fellow Physiologist; to a few as a chief and a friend. The most highly privileged were those who had the honour of working with him or under him, for they could appreciate the true greatness of the man.

Research workers and assistants at University College

George Harley, FRS (1829–96)
Harley was another Scot. Born at Haddington, East Lothian, he became a medical student at Edinburgh in 1846, graduating MD in 1850; he was a contemporary of Burdon Sanderson. He served as resident surgeon and then physician at the Royal Infirmary.

In 1850 he and Burdon Sanderson both went to Paris to study chemistry under Robin, Verdeil and Wurtz, where Harley closely studied the chemistry of urinary constituents such as urea, creatinine, uric acid and hippuric acid. He also attended the lectures of Magendie and Bernard. These were exciting times in Paris; Harley was on the streets during the coup by Napoleon III.

Harley then went to Germany and during 1852–54 studied physiological chemistry with Liebig in Munich, histology with Kollicker and pathology with Virchow in Würzburg and blood gases with Bunsen in Heidelberg.

In 1855, when he went to London, he was very highly qualified as a physiologist and carried the recommendations of his Edinburgh teachers, all known to Sharpey. Sharpey immediately found him a job as Curator of the Anatomical Museum at University College and then as Lecturer in Histology and Practical Physiology, to begin the practical classes to students (p. 86). Harley's demonstrations were edited by a student, G. T. Brown, into a book on histology (1868) with a later edition by Harley himself.

However, in 1859, Harley became Professor of Medical Jurisprudence and a physician at University College Hospital and thereafter his primary work became more and more that of a physician in the hospital and in private practice. He continued for some years to conduct the classes in practical physiology and histology but gave this up in 1866, to be succeeded by Michael Foster. He was a well regarded physician, a Fellow of the Royal College of Physicians, but he held no official positions in that body or in the Administration of University College Hospital. He had been elected FRS in 1865.

Even after retiring from the classes in physiology, Harley kept access to the department as a research worker. In his career he published innumerable papers on a wide range of medical and physiological topics of which the following are a selection: suprarenal glands, strychnine, atropine, substances which give urine its colour, oxygen in blood, pancreatic juice, arrow head poisons. He also did work on asphyxia and chloroform as a member of committees set up by the Royal Medical and Chirurgical Society to investigate suspended animation and the use of chloroform. The experimental work of these committees, and other work, was carried out in the physiological laboratory at University College; Burdon Sanderson was also a member of the committee on asphyxia. Harley was never a member of the Physiological Society.

Amongst this mass of rather superficial research he wrote two books which were regarded as important — *The Urine and its derangements* (1872) and *Diseases of the Liver* (1883).

He wrote also on many non-medical subjects. In 1864 he addressed the British Association on the 'Poisoned arrow heads of savage man', illustrating his talk with specimens from all over the globe. For the Christmas 1879—80 number of *Home* he wrote 'My Ghost Story'; in the letters of the Selborne Society he contributed 'A Blackbird's Widowhood, Wooing and Second Wedding'; in the *Haddington Courier* a series of articles on the buildings of his home town, including the Abbey Church; a book advocating simplified spelling by omitting all double consonants. He wrote on 'Primitive Writings in Sticks and Bones', particularly tracing the writing of the native tribes of Australia, who in fact never did write.

One of his biographers states that he loved to roam over a variety of subjects; he would ponder any matter of interest to him and then record his ideas in print. This was referring to his general writing, but could perhaps also apply to his scientific papers.

Harley's life was punctuated by a series of queer happenings. While still a student in Edinburgh he acted as House Officer to James Simpson in the maternity ward, and single-handed performed a Caesarian section on a woman who died of heart failure, delivering a male child who lived to raise three children; the case was reported by Simpson.

While in Paris Harley thought from the work of Bernard that it would be possible to produce diabetes by stimulating the liver. So he tried the effect in the dog of injecting alcohol, sulphuric acid and chloroform into the portal circulation. He rendered himself diabetic through disordering his digestion by living for three days entirely on asparagus rendered stimulating by pepper and vinegar. The diabetes lasted for fourteen days but he never repeated the experiment.

Another somewhat dangerous experiment on himself was the taking of three to four drops of nitroglycerine, causing his heart rate to fall to forty; he became paralysed from the feet upwards, without losing consciousness. After the administration of ammonia and brandy the alarming effect passed off in about an hour and a half.

At the height of his career in 1864, while he was looking through a microscope, a blood vessel reptured in his left retina. The eye became glaucomatous and when the right eye also became affected he was advised to have the left eye removed. But he decided to try the effect of complete rest of the eyes and lived for nine months in total darkness, not seeing even his own hand in front of his face. He lived in rooms arranged for the purpose and during his darkness dictated to an amanuensis his book on the urine and its derangements. He ultimately recovered vision in both eyes and published in the *Lancet* (1868, i, 158) an account of the curious sensations when he again went into the light. He had no sense of colour and objects were distorted with no sense of perspective, everything seeming very long and his feet far away as in a distorting mirror.

An advocate of cremation, his remains were cremated at Woking and buried, without invitations or flowers, at Kingsbury Old Church.

Sydney Ringer, FRS (1835–1910)

Sydney Ringer was born at Norwich. His father died early leaving three sons, two of whom became extremely wealthy merchants of Shanghai and Japan. Sydney, the second of the brothers, was first apprenticed to a practitioner in Norwich and in 1854 entered University College, London as a medical student. He qualified MRCS and LSA in 1859, MB London in 1860 and proceeded to MD London and MRCP in 1863.

His professional work was as a physician, always in University College Hospital. Beginning as Resident Medical Officer in 1861, he became Assistant Physician (1863), Physician (1866) and on his retirement in 1900 he was given the title of Consulting Physician. Also in 1864–69 he was Assistant Physician to the Children's Hospital at Great Ormond Street. In the hospital he was an excellent bedside teacher, following in the traditions of his immediate predecessors and his own teachers, such as Jenner and Parkes. He also had a considerable consulting practice in Cavendish Place.

In University College Hospital Medical School, he was in succession Professor of Materia Medica and Therapeutics (1862–78), of Medicine (1878–87) and Clinical Medicine (1887–1900). He wrote two books for clinicians. The first was *A Handbook of Therapeutics* first published in 1869; this was a success which went into its fourteenth edition before Ringer retired. The second book, *On the temperature of the body as a means of diagnosis of phthysis*, was first published in 1865 with further versions in 1873 and 1878; effective clinical thermometers were introduced about this time.

After his retirement, Ringer and his wife, who was a Yorkshirewoman from Aldby Park, near York, lived in Lastingham, East Riding. There they contributed to the restoration of the church in memory of a daughter and both are buried in Lastingham.

As set out above, Ringer's life was that of a good physician, busy with his private and hospital work and with clinical teaching. He was a retiring man who revealed little of himself to even his close associates and he and his wife took little part in social activities. A feature of his daily life was punctuality and by this orderliness, combined with driving enthusiasm, he could maintain a second career as physiologist and pharmacologist. Particularly from 1875—95 he used a place which was always at his disposal in the Physiological Laboratory at University College, with Burdon Sanderson and then Schäfer as Jodrell Professor. He became a member of the Physiological Society in 1884, and was elected FRS in 1885.

His daily routine was described by one who was obviously a junior colleague:

He would rise early, dispatch a hasty breakfast at eight, and the next few minutes would see him on his way to the hospital, always on foot, carrying, perhaps, with him some casual co-breakfaster, astonished at the celerity of things. The hospital visit would generally conclude with a quick-change appearance in the physiological laboratory – Ringer the physician transformed into Ringer the pharmacologist. Upon the patiently plodding laboratory assistant these visits came not unlike electric shocks: Fielder! would have been hailed, a tracing taken, various suggestions made, and off he was again on his way back to Cavendish Place and the morning's consulting work. There might be another similar visit to the laboratory in the afternoon.

Ringer himself said that in science he was an amateur, working for his recreation; but, added Schäfer, the sort of amateur who produces better work than many a professional.

His methods were simple. He used simple apparatus such as might be used by students to record contraction of the heart or of muscle or he counted the drops which passed through perfused blood vessels. He used frogs, only rarely mammals, and indeed in his manner of working it would have been difficult to find enough time to work with more complex preparations. In his greatest working period from 1875 to 1895 he published forty-four papers in the *Journal of Physiology*; mostly these papers deal with the action of drugs on tissues of the frog. Beginning in 1882, many papers deal with the effects of salts of Na, K and Ca, particularly on the frog heart, and from this work physiologists have used 'Ringer's solution' to denote a solution of NaCl, KCl, $CaCl_2$ and $NaHCO_3$ in proper proportions which can be used to sustain isolated tissues of the frog. The first introduction of Ca into 'Ringer's fluid' is a classical accident of physiological research. The last paper in Volume III of the *Journal of Physiology* is by Sydney Ringer and the third paper in the next number (Vol. IV) is also by Ringer, with the title 'A further contribution regarding the influence of the different constituents of the blood on the contraction of the heart', and it begins:

After the publication of a paper in the Journal of Physiology, Vol. III, No. 5, entitled 'Concerning the influence exerted by each of the Constituents of the Blood on the Contraction of the Ventricle', I discovered, that the saline solution which I had used had not been prepared with distilled water, but with pipe water supplied by the New River Water Company. As this water contains minute traces of various inorganic substances, I at once tested the action of saline solution made with distilled water and I found that I did not get the effects described in the paper referred to. It is obvious therefore that the effects I had obtained are due to some of the inorganic constituents of the pipe water.

Testing the effect of lime water added to the distilled water saline revealed the effects of Ca.

In trying to assess Ringer's work of almost exactly 100 years ago, it must be remembered that there was then no idea of the intra- and extra-cellular distribution of ions. Ringer merely described the effects he saw in his simple experiments and as such it is the first, now remote, beginnings of much that has followed.

About 1880, some of Ringer's papers had W. Murrell (p. 222) or E. A. Moreshead as collaborators and later H. Sainsbury, D. W. Buxton or A. G. Phear. In the titles of the papers they are usually described as assistant to the Professor of Materia Medica or Medicine.

Edward Emanuel Klein, FRS (1844–1925)
Although Klein had no formal connection with University College, he is noted here because of his close connection with Burdon Sanderson. Emanuel Klein was born at Essek in Slavonia; Edward was added to

his name after he became resident in London. He graduated MD in Vienna; having devoted himself to the newly developing study of microscopic anatomy, in 1868 he was appointed assistant professor to Stricker. In 1869, the New Sydenham Society determined to publish an English translation of Stricker's *Manual of Human and Comparative Histology* and Klein, although he spoke little English, was sent to London by Stricker to make arrangements. There Klein met Huxley, Simon and Burdon Sanderson.

In 1870–71, Sir John Simon, then Medical Officer to the Privy Council (p. 41), was responsible for the use of a grant of £2,000 to investigate infectious diseases, in which Burdon Sanderson was then the most active worker. Both recognised in Klein a skilled microscopist, most able to carry on the work of recognising the particulate nature of contagion as revealed by the microscope. Klein was invited to live in London and to work in the Brown Institute (p. 136) of which he was resident Deputy-Director from 1871 to 1890, with the particular task of studying infectious diseases by microscopic methods following the work of Burdon Sanderson. Some of this work was also carried out in Burdon Sanderson's private laboratory.

Klein's work of this period appeared as excellent Local Government Reports dealing with such topics as sheep-pox, typhoid fever, scarlatina, epidemic diarrhoea, smallpox, cholera, and the bacterial contamination of water. He was a pioneer of bacteriology at the time of Pasteur and before Koch, which was recognised by his election as Fellow of the Royal Society in 1875. He wrote an important early textbook of bacteriology, *Micro-organisms and Disease; an introduction into the Study of Specific Micro-organisms* (1884). In 1890 he opened in Great Russell Street a private school of bacteriology, where he trained many famous men, including Ronald Ross. However, a few years later he was given a laboratory at the top of St Bartholomew's Medical School, where he continued advanced teaching and carried on research work supported by the Local Government Board. In 1902 Bart's recognised his standing as a bacteriologist by appointing him Lecturer in Advanced Bacteriology; with this title he continued to work at Bart's until his retirement about 1914.

Although he must be remembered as a bacteriologist, as an histologist he was associated with the early years of experimental physiology. From his early work with Stricker, his work on infections led to a proper description of the lymphatic system in *The Anatomy of the Lymphatic System*; two books published in 1873 and 1875.

He also wrote the histological section in *A Handbook for the Physiological Laboratory* by Burdon Sanderson, Foster, Brunton and Klein. He produced two major works on histology during his period at the Brown Institute: in 1879, with Noble Smith, a magnificently illustrated *Atlas of Histology* and in 1883 *Elements of Histology*, which was a very successful textbook repeated in several editions.

In 1873 Klein was appointed Lecturer in Histology at St Bartholomew's Medical School and he remained connected with Bart's until his retirement. At first he shared the teaching of anatomy and physiology with Morrant Baker (p. 53) but became sole lecturer from 1884–1900, thereby introducing into Bart's the teaching of experimental physiology as well as good histology. From 1900 to 1902 he shared the teaching of physiology with J. S. Edkins and in 1902, when he was appointed Lecturer in Bacteriology, gave up the teaching to Edkins. When he first began lecturing at Bart's his broken English amused the ribald students.

Klein soon became a naturalised Englishman and in 1877 married an Englishwoman. He was a first-rate chess-player and very fond of music. He came to have some influence in the scientific world of his adopted country, serving on the council of the Royal Society from 1888–1890. He was asked to make official investigations on epidemics and in his partial retirement was employed by the Fishmongers' Company to establish standards for the pollution of shellfish.

Klein was a founder member of the Physiological Society, despite the anger and concern of physiologists about his evidence to the Royal Commission on Animal Experimentation (1875–76). In effect he said that he only used anaesthetics to stop the animal from struggling. Unfortunately Huxley was not present at this session to cross-examine Klein and so get a truer reflection of his experimental practice and his humane nature. Schäfer suggested that as a European, Klein found it difficult to take the Royal Commission seriously; he was in fact fond of animals.

William Marcet, FRS (1828–1900)

William Marcet was born in Geneva. His family were originally Swiss but his grandfather had adopted English nationality; his father was Professor in Physics in Geneva. William lived and worked in England but always kept Switzerland as his second home, particularly spending his summers yachting and mountaineering; the mountains were the inspiration of his important work in physiology. After schooling in

Geneva, William went to Edinburgh in 1846 and qualified MD in 1850. He was thus contemporary in Edinburgh with Burdon Sanderson and Harley. In 1851, like these men, he went to Paris to study chemistry under Verdeil.

In 1853 Marcet began practice in London as a physician, was appointed Assistant Physician at the Westminster Hospital and in 1858, Lecturer in Chemistry and Toxicology. At this time, apparently in a laboratory at his own house, he studied the action of bile and the physiology of fat absorption and he wrote a book, *On the Composition of Food and how it is adulterated*. He was elected FRS in 1857. He was particularly interested in tuberculosis and abandoned his appointments at the Westminster Hospital when he was elected Assistant Physician at the Hospital for Consumption at Brompton in 1867. He did some work on the infectious nature of the sputum of patients. Later he left his London work as a physician to set up practice in the winter in Nice or Cannes, leaving the summer for yachting and mountain-climbing.

He became associated with Burden Sanderson when they were members of a Royal Commission on cattle plague and a commission on the mode of administration of anaesthetics. Both of these bodies used laboratory facilities of University College, London and so Marcet became a worker at University College from this time afterwards, apparently providing his own apparatus − he was wealthy. Marcet was a guest of Burdon Sanderson at the first meeting of the Physiological Society in 1876 and was also a guest on other occasions before he became a member in 1886.

He worked in a laboratory in University College from about 1875 to about 1895, welcomed firstly by Burdon Sanderson and then by Schäfer, carrying on work which was of great importance in the developing science of accurate physiological measurement. He measured respiratory volume, collected expired air and measured its content of CO_2 and O_2, thus demonstrating CO_2 production and O_2 uptake and relating them to give a 'respiratory ratio'. He investigated the effect of cold on respiratory exchanges and he also investigated the effect of altitude. During climbing expeditions he measured respiratory exchanges on Breithorn, Zermatt (13,685 feet), Col Théo dule (10,899 feet), Col du Géant (11,000 feet) and lived for three weeks on the Peak of Tenerife at 8,000 to 13,000 feet. On such expeditions his guide helped carry the apparatus and acted as experimental assistant.

In the laboratory at University College he had designed a carefully balanced spirometer which he used to collect expired air and also to give a record of the respiration. He also erected a human calorimeter, perhaps the first, which he showed to a meeting of the Physiological Society in 1897. By its use Marcet made measurements of heat production and related them to rate of O_2 uptake. This important work was published in *Transactions of the Royal Society* and was presented in the Croonian Lectures to the Royal College of Physicians in 1895 as 'A contribution to the History of Respiration in Man'; this was also the title of a book published later.

In his mountaineering Marcet studied climatic conditions, producing a book on the climatic conditions of *The Principal Southern and Swiss Health Resorts* (1883). He was much interested in and a founder of meteorology, being an early member of the Meteorological Society, and its president in 1888. Also he had studied the effects of alcohol on man, writing a book in 1860 on this subject.

Marcet's most important work was carried out in a laboratory in University College after Schäfer became a professor, and Schäfer became a friend and admirer of the older man, writing an appreciative obituary notice for the Royal Society. When Marcet retired from active work he gave his apparatus to Schäfer, including the human calorimeter. This was a double chamber large enough to contain a seated human subject, with the inner chamber made of well-polished copper and the outer of wood thickly padded with felt. Air was introduced and circulated by electric fans driven by the same power source as lit the laboratory. Heat production was measured by the rate of melting of ice placed within the inner chamber with the subject. In the obituary notice Schäfer said that he had taken the apparatus to Edinburgh where it was then in use.

By the accounts of Schäfer and others Marcet was an attractive companion and a most welcome guest and then member at Physiological Society meetings. He retired to Geneva about 1898 and died at Luxor on a trip along the Nile undertaken for reasons of ill-health.

Frederick James M. Page (1848–1907)
Page was not medically qualified. He obtained BSc (London) in 1869 and became FIC in 1871; he was a student of the Royal School of Mines.

In 1878 he was assistant to Burdon Sanderson at University College, London and was elected a member of the Physiological Society. The teaching of physiological chemistry was delegated to Page and he also assisted Burdon Sanderson in experiments. In 1879 Page published a paper in the *Journal of Physiology* on the CO_2 production of dogs at varying temperatures, reporting work which had been carried out at the Brown Institute. In 1880 and 1884 there were papers with Burdon Sanderson on the excitability of the frog heart (p. 144). Over these years Page also gave several demonstrations to the Physiological Society, one of them on behalf of Ringer.

In 1885 Page was appointed Lecturer in Chemistry at the London Hospital Medical College and he there also gave classes in practical physics. By then these subjects had grown too complex to be taught by young surgeons or physicians and the hospitals therefore paid specialists to teach them. Page was apparently a modest and retiring man but well liked by his students. He wrote two books for students – *A manual of chemistry, inorganic and organic* which went into a third edition and *Elements of physics for medical students*, published in 1907 shortly before his death.

In 1907 a minute of the Physiological Society records the death of Page, an early member, and expresses sympathy to his widow.

References

Bellor, H. (1929). *University College, London 1826–1926. University of London Press.*

Lady Burdon Sanderson and J. S. and E. S. Haldane (1911). Sir John Burdon Sanderson – a memoir. Oxford: Clarendon Press.

Klein, E., Burdon Sanderson, J., Foster, M. and Brunton, T. L. (1873). *Handbook for the Physiological Laboratory*, edited by J. Burdon Sanderson. 2 vols. London: Churchill.

Schäfer, E. A. (ed.) (1898). *Textbook of Physiology*. 2 vols. Edinburgh and London: Pentland.

Whitteridge, D. (1983). 'The origin of the *Quarterly Journal of Experimental Physiology*', *Quarterly Journal of Experimental Physiology*, **68**, 521–3.

Biographies

Burdon Sanderson: *Proc. Roy. Soc. London B* (1906–7), **79**, iii–xviii (by Gotch) (P).
British Medical Journal (1905), ii, 1481–92 (P).
Lady Burdon Sanderson (1911).

Harley: *Proc. Roy. Soc. London* (1897), **61**, v–x.
Lancet (1896), ii, 1330–3 (P).

Klein: *Proc. Roy. Soc. London B* (1925), **98**, xxv–xxix.
                        Lancet (1925), i, 411–12.

Marcet: *Proc. Roy. Soc. London* (1905), **75**, 165–9 (P)
                        (by Schäfer).
                        Lancet (1900), i, 811–12.

Osler: *British Medical Journal* (1920), i, 30–3 (P).

Page: *British Medical Journal* (1907), ii, 557.

Ringer: *Proc. Roy. Soc. London B* (1911), **84**, i–iii (P)
                        (by Schäfer).
                        British Medical Journal (1910), ii, 1384–6.

Sharpey-Schafer: *Obituary Notices of Fellows of the Royal Society*,
                        London (1935), 1, 401–7 (P) by L. Hill.
                        Lancet (1935), i, 843–5 (P).

11

Physiologists at Cambridge

The developing department

In listing courses for medical students at Cambridge for the session 1870–71, the *Lancet* of 10 September 1870 shows lectures in 'Anatomy and Physiology' and in 'Anatomy, Descriptive and Surgical', each three per week, given by Professor Humphry. There was no practical course in physiology and there was no separate subject of physiology in either medical examinations or the Natural Sciences Tripos.

From 1871 until 1880, the *Lancet* lists the same lectures by Professor Humphry but there are now three classes per week in Practical Physiology, given by Dr Foster. Also, physiology became listed as a separate subject for medical examinations and in the Natural Sciences Tripos. The number of students who took the practical classes was not large; even in 1880 there were only thirty-two students in the class list for Natural Sciences Tripos, of whom perhaps one third would have taken physiology.

The instrument of change was the appointment of Michael Foster as Praelector in Physiology at Trinity College. Langley (1906) recalled: 'The University gave to the Trinity Praelector a room for use both as a lecture theatre and as a laboratory. Trinity College supplied the simple furniture, some tables, some bottles and reagents, a few microscopes and so on. Three tables put together at the top of the room served to mark off a space which served as the Lecturer's platform'. Foster and his assistants first taught and carried on research in these makeshift conditions until, with physiology and biology becoming recognised subjects and with the increasing status of medical teaching in Cambridge, the University built a biological laboratory, which was opened in 1878. This was soon further extended to give more room for physiology, botany and zoology and also for human anatomy. This building was on the east side of Downing Street, where the Cavendish Laboratory had been built in 1874. It was originally the site of a priory and later the Botanical Gardens, which were moved to their present site in 1853, thus freeing the central site for new buildings for the growing scientific faculties.

The teaching staff in 1871 consisted solely of Michael Foster, assisted by

H. Newell Martin who came with him from University College, being awarded a Scholarship by Christ's College and later a Fellowship. He stayed only until 1876. Soon Foster had enticed some of his students to begin careers as physiologists. By 1875, Gaskell and Langley were assisting in teaching and carrying out research work; others less well known to modern physiologists were Lea and Dew-Smith. Gaskell (1907) wrote: 'Such a growth was only possible, owing to the endowments of the colleges; for these young and enthusiastic teachers who gave their whole time to Foster and his work, with neither appointment nor salary from the University, could not have done so but for the fellowship system and the recognition by their respective colleges that they were doing work worthy of support.' Gaskell and Langley became physiologists of major importance – Langley succeeding Foster as professor in Cambridge.

The physiology taught by Foster is shown in his *Textbook of Physiology*, which he wrote after he went to Cambridge, the first edition being published in 1877. This book was a landmark in the writing of modern physiology (p. 31). Foster's practical classes are set out in a laboratory manual which he produced for his practical classes, with Langley as assistant editor. *A Course of Elementary Practical Physiology* was published in 1876; experiments to be done by the students were largely chemical and histological, with some simple frog experiments and some demonstrations set up by the teacher.

Foster saw physiology as having importance in two ways. Firstly, it was a subject for the medical curriculum. After a period of almost complete lapse, medical education in Cambridge was being revived by a surgeon and anatomist, Humphry, and a physician, George Paget.

George Murray Humphry, FRS (1820–96) was in 1870 Professor of Human Anatomy. Primarily he was a surgeon of the old school with a large practice in the Cambridge district. Born in Sudbury, after apprenticeship with a surgeon in Norwich, he entered St Bartholomew's Hospital in 1839 as a student and came under the influence of James Paget (p. 49). He qualified in 1842 and, by the recommendation of James Paget, was appointed surgeon to Addenbrooke's Hospital, aged only twenty-two. He there began his life-long colleague relationship with George Paget, physician and brother of James.

In 1847, Clark, the Professor of Anatomy, asked Humphry to give lectures in anatomy and Humphry then entered the University as a fellow-commoner of Downing College, graduating MB in 1852 and MD in 1856. Clark resigned in 1866 and two professors were appointed; Humphry as Professor of Human Anatomy and Newton (see below) as Professor of Zoology and Comparative Anatomy.

In 1867, with Turner of Edinburgh, Humphry founded the *Journal of Anatomy and Physiology* and was its chief editor. He continued to be an active surgeon and teacher of surgery and became a great public figure, being knighted in 1891. He was also a great anatomist, his most important book being *The Human Skeleton* (1858) and he was elected a Fellow of the Royal Society in 1859. He was sufficiently interested in physiology to become in 1876 a founder member of the Physiological Society.

George Edward Paget, FRS (1809–92) was an elder brother of James Paget (p. 49). In his boyhood the family at Yarmouth was sufficiently wealthy to send George to Charterhouse and then to Caius College, Cambridge. He took his BA in 1831 as eighth wrangler, then studied medicine at St Bartholomew's Hospital and in 1833 returned to Cambridge to take his MB degree and start practice. In 1839 he was appointed physician at Addenbrooke's Hospital, which post he retained for forty-five years; in 1872 he became Regius Professor of Physic and was elected FRS in 1873. Much honoured as a physician, he was knighted in 1885.

Humphry and Paget combined to promote the scientific aspects of medicine and were an influence behind the appointment of a Praelector in Physiology at Trinity in 1870 and the establishment in 1883 and 1884 of University Chairs in Physiology and Pathology. Certainly, in Humphry and Paget, Foster had two powerful patrons in the introduction of practical physiology into the medical curriculum.

Secondly, Foster regarded physiology as one of the biological sciences, which must go hand in hand with botany and zoology; he continually sought to establish a School of Biology in Cambridge. In this he was inspired by his association with Huxley and continually sought to ensure that practical classes should be the mode of teaching in all of the subjects of the new biology; by 1880 botany and zoology also had laboratories. When he first went to Cambridge, Foster lectured on embryology and elementary biology as well as physiology until by 1880 these lectures were taken on by his pupils: Vines, later Professor of Botany in Oxford; Balfour, later Professor of Animal Physiology in Cambridge and Sedgwick, later Reader in Animal Morphology; Foster then lectured on physiology only. However, the important point is that by 1880 there was in Cambridge an effective School of Biology, largely due to Foster's influence.

Foster's practical classes were in addition to the official lectures given by the professors. The Professor of Botany was

Charles Cardale Babington, FRS (1808–95) who was a noted systematic botanist. He had made many visits to all parts of the British Isles and produced in 1843 a *Manual of British Botany* which over many editions had a great influence upon field botany. He was elected FRS in 1851. He was also an archaeologist. Babington's lectures were mainly on anatomical lines and, despite dwindling classes, he had little sympathy with histological or physiological detail and there was no laboratory work. For some time before his death, his health failed but he retained his chair, giving half his salary to a deputy.

Alfred Newton (1829–1907) was Professor of Zoology and Comparative Anatomy, the Chair having been created for him in 1866. Newton was a great authority on birds, a knowledge acquired on extensive travels in Lapland, Iceland, the West Indies, North America and in the British Isles. He spent his holidays in the bays and sea-lochs of Scotland on a friend's steam-yacht. These field studies were carried out despite a serious hip disease which required the use of two sticks. From them he wrote a classical book, *A Dictionary of Birds* (1893–96). In Cambridge the most effective part of his teaching was in his rooms in the Old Lodge of Magdalene College surrounded by his books. Particularly he was at home to scientists and students on Sunday evenings, when Charles Kingsley sometimes attended. He held no practical classes.

By the nature of these professorial colleagues Foster was able to establish a school of biology of a newer type which created its own young men of the new more advanced thinking, such as Vines in botany and Balfour in zoology; Foster's essential contribution was the introduction of practical classes into botany and zoology following the lead of Huxley.

Research began as soon as Foster went to Cambridge. At first papers were published in the *Journal of Anatomy and Physiology*, the publication founded by Humphry of which Foster became co-editor. The issue of November 1873 contained papers from the Physiological Laboratory in the University of Cambridge by H. Newell Martin (Scholar of Christ's College), M. Foster (Praelector in Physiology in Trinity College), A. G. Dew-Smith (Trinity College), C. J. F. Yule (Scholar of St John's College). In 1873, 1876 and 1877, Foster edited *Studies from the Physiological Laboratory in the University of Cambridge* containing papers by Langley, Marshall, Martin, Dew-Smith, Vines and Gaskell. In a preface, Foster thanked the University 'for having permitted me, a simple College Lecturer, to occupy, at some inconvenience I fear to others, the two University rooms in which my lectures are given, the practical teaching of my class conducted and the physiological work carried on. I have presumed on their kindness and ventured to call these rooms the Physiological Laboratory of the University of Cambridge.'

When in 1878 Foster founded the *Journal of Physiology*, the first three volumes (1878–82) contained twenty-two papers from the Physiological Laboratory in the University of Cambridge by nine authors, mostly described as scholars or fellows of colleges. One of the authors was C. S. Roy, the first G. H. Lewes Scholar, and two were visiting Americans. The chief contributors from Cambridge were Gaskell (four papers) and Langley (six papers).

At first Foster himself did some research. In London before 1870 he had published four short papers on several topics, including the effects of electrical stimulation of the frog ventricle. In Cambridge he continued to work on the effects of stimulation and temperature on the heart and reflexes of the snail and frog. None of this work turned out to be particularly important and after 1876 the papers from Cambridge were the work of his pupils. The topics of the twenty-two papers from Cambridge were wide-ranging – e.g. North's work on urea excretion by man, Langley's experiments on salivary secretion at this time largely histological, the effects of stimulation of nerves to the heart and blood vessels of muscles by Gaskell, the proteid contents of seeds by Vines. Some of the experiments of Gaskell and Langley were on cats, dogs or rabbits anaesthetised with morphia and ether or chloroform.

An experimental laboratory needs apparatus. In the 1870s laboratory research in England was much handicapped by the difficulty of getting any special instrument made for any particular research. Dew-Smith (p. 181), one of Foster's early pupils and also his supporter in setting up the *Journal of Physiology* (p. 262), saw this need and set up in Cambridge a workshop expressly to turn out anything required in scientific laboratories as quickly as possible. This workshop was enlarged by Dew-Smith into the Cambridge Scientific Instrument Company of which Horace Darwin later became sole director. *Horace Darwin* (1851–1928), youngest son of Charles Darwin, was something of a genius in designing experimental apparatus and methods.

The atmosphere of Foster's laboratory has been described by Gaskell, one of his first Cambridge pupils:

'Foster held very strong views as to the proper method of teaching physiology to students at the beginning of their medical study. He held it to be a mistake to demonstrate during the lecture, and insisted that practical work, carried on by the student himself, illustrative of the facts on which the lecture was based, must immediately follow the lecture. (In this he was following principles used by Huxley.) The physiology of each organ must be dealt with as a whole in the lecture, and the practical work must be so arranged as to bring home to the student all of the points of each lecture at the time, and not to be regardless of the lecture, as must be the case if the practical work is departmental while the lecture course is general. His ideal laboratory would be of sufficient size to provide each student with his own working place, both in the histological and in the chemical department at the same time. He also –

and this was one of the great reasons of his success − encouraged his students at the very earliest moment to engage in some original research, and then persuaded them to give a few lectures of an advanced character upon the subject on which they were working; for, as he said, there is no way of discovering gaps in your knowledge of a subject better than lecturing on it.

Once started on a research topic, a student was often persuaded into a scientific career. 'By his (Foster's) earnestness, his lovable charm of persuasion, the conviction was gradually born in on his pupils that the particular line of research on which each was engaged was the one thing in life worth doing, and that the only place to do it was in Cambridge by Foster's side. As Foster used to say "the true man of science must feel with respect to his own research that in this way only lies salvation".'

Foster's method was eminently successful in the surroundings of Cambridge, where students were rather few in number and could thus be well known to their teachers and where a man's first years in research could be supported by his college. Even today a large part of Foster's system persists in the Part II of the Natural Science Tripos and in the third year of BSc Honours courses of other Universities. A small number of students there become closely known to the staff, are given 'projects' which can lead towards a research topic and a selected few are taken on to research grants and PhD studies.

The professor

Sir Michael Foster, FRS (1836–1907)

Michael Foster was born at Huntingdon, where his father was a surgeon and a prominent citizen. Michael went to University College School, where he distinguished himself in classics. He then took his London BA degree brilliantly but could not seek a classical scholarship at Cambridge, because the Fosters were staunch Baptists. It was therefore decided that he should follow his father's profession and in 1854 he entered University College, London as a medical student. He gained the gold medal in anatomy and physiology, took his MB London in 1858 and MD in the following year. As a brilliant student he came to Sharpey's notice and was much attracted by Sharpey's teaching. He was not, however, immediately drawn into physiology as a career, although he did join the British Association and a paper on 'The effects produced by Freezing on the Physiological Properties of Muscle' appeared in the Proceedings of the Royal Society in 1859–60.

In 1859–60 he continued his medical studies in Paris. Much later, in 1899, he wrote a memoir of Claude Bernard in the 'Masters of

Medicine' series and in the foreword declared that he never met Bernard nor attended any of his lecturers, although he had great reverence for Bernard and his teaching.

Anxiety that he might develop tuberculosis led to his spending 1860 as ship's surgeon on the *SS Union*, building a lighthouse in the Red Sea on the Asaruf rock opposite Mount Sinai, where he studied some marine animals. From 1861 he was in practice in Huntingdon with his father but remained interested in biology, particularly as a member of the British Association.

Foster's real work began in 1867 when he was invited by Sharpey to assist and then succeed Harley (p. 86) as lecturer in charge of the practical classes at University College; in 1869 his title was changed to Assistant Professor. He was quickly recognised and in 1869 followed Huxley as Fullerian Professor at the Royal Institution. In Huxley's course of elementary biology at South Kensington, the demonstrators in the practical class included Foster, Lankester and Rutherford.

The short period that Foster was at University College was important to his development. Sharpey's practical classes were well established and unique in that the students, in addition to microscopic work and chemistry, also carried out some experiments; Foster later described what was done in these classes (p. 87). Also Foster became associated with Huxley and his practical classes and this association led Foster to relate physiology to a general biology, so that when he went to Cambridge his aims included placing physiology at the centre of a group of biological sciences.

Wider opportunity came in 1870 when Trinity College, Cambridge decided to have a Lecturer in Physiology. The Trinity fellows who most influenced this decision seem to have been W. G. Clark, Public Orator in the University and Coutts Trotter, later Vice-master. According to Gaskell (1907) and Sharpey-Schafer (1927) they were advised by Huxley, G. H. Lewes and George Eliot. Clark certainly was a friend of Lewes and George Eliot but their role in the matter is uncertain, that of Huxley more likely. Thus in 1870 Foster was appointed Praelector in Physiology at Trinity College, Cambridge. As described in the previous pages, from this appointment Foster built a School of Physiology in Cambridge, which within ten years probably became the best in the world.

Foster went to Cambridge with a college appointment only; he had no appointment in the University. The University conferred on him the honorary degree of Master of Arts, but this gave him no administrative status. He was ineligible for election to Boards of Studies and could only make his voice heard in the University through his friends. In 1883, consequent on the report of the Royal Commission, a Professorship of Physiology was founded in Cambridge (and in Oxford, p.144) and the Trinity Praelector became a university professor without any practical change in his duties and without much extra expense to the University; the full degree of MA was then conferred on Foster and he could speak for himself in university affairs. However he never took much part in routine administration. His keen interest was in the establishment of new studies, particularly, for example, in the inception of the Department of Agriculture. He continued until 1900 with his teaching of students and encouragement of research. Also he wrote successive editions of his textbook (p.31).

Outside the University he was active in matters related to physiology, beginning with the formation of the Physiological Society in 1876 (p.262), of which he was an active founder, presided at the inaugural dinner and was thereafter a leading member. At a breakfast party given by Yeo to Goltz, Foster, Heidenhain and Kronecker during the 1881 Congress of Medicine in London, the idea was mooted that there should be an International Congress of Physiologists; it was largely Foster who worked on this plan to achieve the First International Congress in Basle in 1889. He was President of the 1898 Congress in Cambridge and in 1901 at Turin was made Honorary Perpetual President. His purpose always was to make the congress demonstrational rather than the reading of papers and to make the social proceedings cheap and informal.

Also, from its inception in 1878, Foster was the sole effective editor of the *Journal of Physiology* (p.262), until 1894, when the *Journal* was in financial difficulty, it was bought by Langley. Thereafter Langley was its working editor although Foster's name remained on the title page as co-editor.

His interest in biology made Foster an active member of the British Association, which he first joined in 1859. He was secretary of its biological section in 1867, Deputy President of the sub-section of anatomy and physiology in 1870, Joint General Secretary in 1872–1876, President of the physiology section in 1897 and President of the Association in 1899.

Foster was elected Fellow of the Royal Society in 1872 and in 1881 succeeded Huxley as one of its secretaries until he resigned in 1903. His individual contribution to the society was to establish close, confidential and frequent relations between the Royal Society and Government departments; the role of the Royal Society as an advisor to Government departments was a new function, to which some fellows were opposed, but he carried his policy through. During his secretaryship he took part in the organisation and establishment of the Meteorological Office and the National Physics Laboratory. He also urged on the government the advisability of grants in aid of scientific investigations, such as the scientific investigations of coral reefs, earthquakes and the exploration of little known countries. He also promoted a Royal Society catalogue of scientific papers leading to an international catalogue, this being expression of his maxim that 'the next best thing to knowing a thing is to know where to find it'.

Like other prominent men of the 1890s Foster was a working member of commissions and enquiries — on the prevention of malaria and tropical diseases, vaccination, disposal of sewage, tuberculosis. He was a member of the commission for the reorganisation of the University of London.

Foster was knighted in 1899 at the same time as Burdon Sanderson. The Physiological Society gave a dinner in their honour, presided over by Schäfer and this must be rated as one of the great occasions in the history of the Society, of which all three were founders.

From 1900 to 1905 he represented the University of London in the House of Commons. By family background and by choice Foster was a Liberal at a time when the Liberal Party was divided on several issues. At first he was able to support the existing Conservative government but later crossed the house to sit with the opposition, chiefly because he could not support their Education Bill. He was again a candidate at the election in 1906 but was defeated by a narrow majority.

Increasing commitments outside Cambridge led to his asking in 1900 for a deputy to perform the duties of the professorship and Langley was appointed Deputy Professor. Foster resigned his professorship in 1903 and Langley then succeeded him.

In 1903 Foster also resigned as Secretary of the Royal Society. In 1904 he suffered some ill-health but recovered to activity and cheerfulness. On Monday 28 January 1907 he gave a characteristically humorous speech in London at a meeting of the British Science Guild, was taken ill in the evening and died that night from pneumothorax caused by the bursting of an oesophageal ulcer; he was attended by his friend Rose Bradford. He was buried at Huntingdon; a memorial service was held in the Chapel of Trinity College.

Outside his scientific work, Foster delighted in cricket and in gardening. At University College School he was captain of the eleven. In Cambridge when at first there were few students, he inaugurated an annual cricket match between the staff and the students. Later, when the department became larger, this was replaced by a match between teachers and assistants in the biological laboratories. The match was played on a field belonging to Foster, who was always captain of his side and played regularly until 1895.

Gardening was his chief relaxation and he did much of the manual work himself. His house was situated on a chalk-slope of the Gog-Magog hills about four miles from Cambridge and there he had large beds of cyclamen, anemones, daffodils, iris, etc. It was a magnificent flower display in which he entertained his friends. He had a special fancy for irises, of which he had a wonderful collection and of which he manufactured many new hybrids, requiring much patience because in many cases the seeds did not yield flowering plants for five years or more. There are also Shelford hybrids of *eremurus*. He had a wide knowledge of horticulture, recognised by his appointment as Chairman of the Departmental Committee to report on the Botanical Collections at Kew and the British Museum and he wrote many articles for horticultural journals, chiefly about irises.

His character can be summarised by compounding from the last paragraphs of obituary notices by Gaskell and Langley:

Sir Michael Foster's varied work in life was only made possible by his sincere and genuine nature. He was a man of large aims, and generous enthusiasms, of strong initiative and unusual powers of inducing others to see as he did. There have been many greater scientific men than Foster. It is hardly too much to say that no man ever devoted himself more wholeheartedly to science and few have done it greater service.

Foster's early students

Henry Newell Martin, FRS (1848–93)
Martin was born in Ireland, the son of a Congregational minister, who later became a schoolmaster. He was educated at home and was able to matriculate in the University of London before he was sixteen. He was apprenticed to Dr McDonagh in Hampstead Road, with the

understanding that his duties would not preclude study at the medical school in nearby University College, London.

At University College in 1868 he introduced himself to Michael Foster, then conducting the practical physiology classes, as Foster later recalled:

My lectures and classes on experimental physiology were absolutely voluntary and only the better students were willing to give up the time needed to get a more thorough grasp of physiology. Well, I appointed a time to see the few who wished to spend some time in this new study, this study of luxury, and there came to me a boy, nothing more than a boy, who said: 'I am very sorry, sir; I would like to take your course if I could, but you see my parents are not very well off, and I get my board and lodging by living with a doctor close by. I have, in return for my board, to dispense all the doctor's medicines, and that dispensing takes me from two to five; now your lectures begin at four. I cannot come for the first hour. I will work hard and try to make up the lost time.' I said 'certainly, certainly'. So he came in, came in regularly late. He came in regularly at five o'clock, and he worked with such purpose that in the examination which I had at the end of the course I awarded him the prize. Well his name was Henry Newell Martin, and I was so struck with him that I asked him to assist me in my course, and he became my demonstrator (cited by Sewall, 1911).

Thus in 1868−70 Martin was Demonstrator in Foster's classes at University College and he became known to Huxley. He also demonstrated in the summer in a class in histology given by Humphry, Professor of Anatomy and Physiology at Cambridge. In 1870 he was awarded a scholarship at Christ's College, Cambridge, a happy coincidence with Foster's appointment as Praelector at Trinity College. Foster continued the story:

After we had been at University College together, I think two or three years, Martin carrying on his studies and at the same time helping me, he came one day to see me in great trouble because he could not make up his mind. He obtained what they call a scholarship at Christ's College at Cambridge and he could not make up his mind to accept and go there. He said he didn't want to leave me. But I was able to tell him what nobody else knew at that time, that in the October in which his scholarship would take him to Cambridge, I was going to Cambridge too, having been invited to lecture there.

In Cambridge Martin continued as a brilliant student, gaining first place in the Natural Sciences Tripos of 1873, taking also BSc and MB of London and he proceeded to become the first Cambridge student to take DSc in physiology. In 1874 his brilliance was recognised by election to a Fellowship of Christ's College.

In Cambridge, he at once undertook to continue as Foster's

Demonstrator, whose right hand he continued to be for the next five years. Foster recalled that Martin, by his nature, had also an indirect effect in promoting physiology in Cambridge:

His energy and talents, and especially his personal qualities, did much to advance and render popular the then growing School of Natural Science in the University. At that time there was, perhaps, a tendency on the part of the undergraduate to depreciate natural, and, especially, biological science, and to regard it as something not quite academical. Martin, by his bright ways, won among his fellows sympathy for this line of study, and showed them, by entering into all their pursuits (he became, for instance, President of the Union and Captain of the Volunteers) that the natural science student was in no respects inferior to the others.

Over the years from 1868–75, Martin was involved in the growth of practical classes in physiology in association with Foster, and in general biology with Huxley. Under Huxley's supervision he prepared a textbook for the course in general biology, which was published in 1876 under the title *Practical Biology* by Huxley and Martin.

By his election as a Fellow of Christ's College, Martin had an assured future in Cambridge. However, in 1876, Johns Hopkins University was being founded in Baltimore. It was at first to have no medical school but a Professor of Biology was to be appointed. Martin was recommended by Foster and Huxley and so in 1876 he moved to Baltimore and his career as a physiologist was thereafter fulfilled in America. He was one of the American co-editors of the *Journal of Physiology* when it was founded in 1878. In Baltimore he was a major influence towards the establishment of a medical school in 1893. He took part in the preliminary organisation, but his health failed and he resigned a few months before the medical school was opened; its first Professor of Physiology was his pupil, W.H. Howell. By his close association with Foster in 1876, Martin was naturally a founder member of the Physiological Society and maintained his membership although he immediately went to America.

Soon after he arrived in America, Martin married the widow of General Program, a Confederate Officer, who was a celebrated beauty. Together they made their home into a social and conference centre to the great pleasure and benefit of students and colleagues. Their social life ended with the death of his wife in 1892 and his own ill-health.

After his resignation in 1893, he returned to England, with perhaps some hope that his health would allow him to resume physiological research in Cambridge. But his illness progressed and within a few months, while staying for his health's sake in Burley-in-Wharfedale, Yorkshire, he died from a sudden severe haemorrhage.

The foregoing account is drawn from Foster's obituary notice for the Royal Society and emphasises Martin's career in Cambridge. Fye (1985) has studied Martin's years at Johns Hopkins, particularly his struggle to get the Huxley–Foster type general biology accepted, with physiology as a central subject. Fye also discusses the illness from which Martin suffered, which was polyneuritis due to alcohol; Martin was an admitted alcoholic who attempted to control his addiction but never succeeded for long. It is not clear whether this was known to Foster.

Martin was a great experimentalist, who did pioneering work in two fields (Breathnach, 1969). From 1880, papers in the *Journal of Physiology* describe the mechanisms and nervous control of breathing, particularly the roles of external and internal intercostal muscles. After about 1885, he began publishing papers on the isolated mammalian heart in which he produced what is in effect a 'heart–lung' preparation, with which he preceded by about twenty years the conclusions reached by Starling. Knowlton and Starling (1912) cited Martin's Croonian Lecture of 1883 as a starting point but the work of Martin has been so submerged by the later work of Starling as to be virtually unknown to present-day physiologists. He was elected FRS in 1885, having given a Croonian Lecture in 1883.

Walter Holbrook Gaskell, FRS (1847–1914)
Gaskell's father was a barrister in London, but Walter Holbrook was born in Naples, where the family was passing the winter. Their home was in Highgate and Walter went to school there. He entered Trinity College, Cambridge in 1865, was elected to a scholarship of Trinity and proceeded to Bachelor of Arts in 1869 in the Mathematical Tripos. He then began to study for a medical career and in 1870 attended the first classes of Michael Foster. Although Gaskell went to University College Hospital for clinical work, Michael Foster had already noticed him and in 1874 Gaskell accepted the idea of a career in physiological research. Instead of proceeding with medical studies, Gaskell went in 1874 to Leipzig to work with Ludwig.

Returning to Cambridge in 1875, Gaskell married and settled in Grantchester and resumed research work in Foster's laboratory. In 1878 he took the degree of MD by thesis, but he never practised medicine. In 1883 he was appointed University Lecturer in Physiology and took a more formal part in the teaching, in parallel with Langley who was also appointed lecturer at about the same time. Of Gaskell's lecturing it was said that he was incisive and spoke on controversial points with a half-suppressed enthusiasm which was eminently infectious. Both in his teaching and general outlook on physiology, there was strong contrast between Gaskell and Langley; their effects on a student as described by Dale are cited on p. 179.

Gaskell was elected a fellow of Trinity Hall in 1889 and also Praelector in Physiology at that college. He continued in these posts until he died from a cerebral haemorrhage in 1914, immediately after revising the last proof sheets of his small book, *The Involuntary Nervous System*.

A founder member of the Physiological Society in 1876, Gaskell took full part in its activities and served as treasurer from 1884 until 1896. He was the second treasurer, succeeding Romanes (p. 249) and was succeeded by Waller (p. 215).

In his undergraduate days he took an ordinary part in rowing, cricket, tennis and swimming but was not much attracted by active exercise. Later he was fond of a rubber at whist. His main hobby became gardening. On the hillside at Shelford, almost opposite Michael Foster, he built a house in fifteen acres of sloping hillside which he turned into a terraced garden, where he entertained Cambridge residents and greeted visiting physiologists.

Gaskell from 1875 until about 1890 was an experimental physiologist, firstly studying the heartbeat and particularly deciding that it was of muscular, not nervous, origin. The rhythm of the heart was his subject in a Croonian lecture of 1881 and he was elected FRS in 1882. Also, in 1898, he wrote the chapter 'The Contraction of Cardiac Muscle' in Schäfer's *Textbook of Physiology* (p. 149). Next he studied the morphology and the effects of the nerves to the heart, blood vessels and viscera. In this he was preceding the work of Langley and, although they never worked in collaboration, it is not easy to distinguish where Gaskell's work ended and Langley's began. The difficulty lies partly in Gaskell's excellent book *The Involuntary Nervous System* which was written in 1913 and presents Gaskell's ideas as they had developed some twenty years after he had ceased to work on the topic.

From the morphological aspects of his work, Gaskell was led to enquire into the origin of vertebrates, and reached a theory that vertebrates are descended from some Crustacean-like ancestor. His book *The Origin of Vertebrates* (1908) was too remote from accepted ideas to make much impression.

Basically Gaskell was an accurate and careful observer and experimenter. 'But the bent of his mind lay in the direction of generalisation. A fact once definitely ascertained was never viewed by him as an isolated phenomenon; it was used as a basis for formulating some general rule.' This was the conclusion of Langley in an obituary notice and herein lay the difference between Gaskell and Langley.

John Newport Langley, FRS (1852–1925)

Langley was born at Newbury, son of a schoolmaster and was educated at home and at Exeter Grammar School, of which his uncle was headmaster. He entered St John's College, Cambridge in 1871, reading mathematics, history and literary subjects with the intention of entering the Civil Service. However during his second year he gave up other subjects to read Natural Science. He attended Foster's lectures and practical classes in May 1873, and thereafter he was always a physiologist in Cambridge.

Foster immediately recognised the quality in Langley – 'From the very first I marked him as a man of whom something was to be made' – and Langley himself described what then happened: 'Foster led a considerable number of his early pupils to a scientific career. He first aroused an interest in scientific problems and then, sometimes gradually, sometimes suddenly, suggested that there was no better course in life than that of trying to solve them.' So Langley was in 1873 elected to a scholarship at St John's College and in December 1874 was placed First Class in the Natural Science Tripos. He had already begun to help Newell Martin in the practical classes and, having taken his Bachelor of Arts degree in 1875, he became Demonstrator in succession to Newell Martin. In 1877 he was elected to a Fellowship of Trinity College and Trinity was thereafter his College. He kept in rooms in Whewell's Court.

Langley never deviated from his life's work, even to obtain a medical degree; he was the first great English physiologist who was not medically qualified. Apart from a short period in 1874 with Heidenhain in Breslau, he always worked in Cambridge. In 1884 he became a permanent Lecturer in Trinity College and also University Lecturer in Physiology; in 1900 he was appointed deputy professor to Foster and in 1903 succeeded to the Chair of Physiology. He died in that office in November 1925 after only a few days' illness.

Langley's whole activity was in his experimental work and teaching. He undertook no administrative work. His experimental work was soon recognised by election Fellow of the Royal Society in 1883; he served on the council for two periods and was vice-president in 1905. His work also earned him medals and lectureships from this and other institutions and honorary degrees and membership was conferred on him by many universities and institutes at home and abroad.

As Demonstrator from 1875 to 1884, Langley was responsible for the practical classes. In 1875 *A Course of elementary Practical Physiology* was published by Foster with the assistance of Langley; the fourth edition in 1880 was jointly by Foster and Langley. Later he was lecturing three times a week to elementary students and also gave classes for more advanced students and continued with about this load of teaching throughout his professorship. Joseph Barcroft (*Nature*, 1925, **116**, 872–3) recalled that his lectures were mines of information but often difficult to follow and that individual students derived very different amounts of benefit from them. As a Demonstrator it was a matter of conscience to Langley to demonstrate throughout each class period and to get to know each man individually; it was then that his greatness was conveyed to the student.

Despite enlargement in 1890, the old laboratory became quite inadequate and in 1910 the Drapers Company of London offered to build a new School of Physiology at Cambridge. Langley gave close attention to the design and equipment of this new building on the Downing Street site and it was opened in June 1914, but had to wait until after the war for its full use.

Langley was a founder member of the Physiological Society, often gave demonstrations and presided at meetings held in Cambridge. His greatest service to the Society was through the *Journal of Physiology*. Founded by Foster in 1878, in 1894 the *Journal* was in debt and threatened with extinction. Langley paid off the debt and so became the owner of the *Journal*. Although Foster's name remained as co-editor, Langley became the sole effective editor, and he was determined to ensure high quality in published papers. To Langley, the acceptance of a paper was only the beginning of his editorial task. He thought it due to science that the paper should be cast in the most effective form and reduced to the least possible size. This meant correction, often heavy correction, commonly suggested in detail by himself, and it involved the exchange of views with sensitive and sometimes irascible authors. But he made the *Journal* unsurpassed in the high standards of the papers within it. There is permanent relevance in what Langley said in a presidential address to the Physiological section of the British Association in 1899:

Those who have occasion to enter into the depths of what is oddly, if generously, called the literature of a scientific subject, alone know the difficulty of emerging with an unsoured disposition. The multitudinous facts presented by each corner of Nature form in large part the scientific man's burden today, and restrict him more and more, willy-nilly, to a narrower and narrower specialism. But that is not the whole of his burden. Much that he is forced to read consists of records of defective experiments, confused statements of results, wearisome description of detail, and unnecessarily protracted discussion of unnecessary hypotheses. The publication of such matter is a serious injury to the man of science; it absorbs the scanty funds of his libraries, and steals away his poor hours of leisure.

The high standard of the *Journal of Physiology* is Langley's bequest to physiologists and something of his standards has been maintained by subsequent editorial boards. But on the other hand, one of the factors behind the establishment in 1908 of the *Quarterly Journal of Experimental Physiology* (p. 149) was the objection of authors to having their papers 'Langley-ised'.

Langley's important research work was in two phases. He first studied secretion, particularly by the submaxillary glands, beginning in 1874 from a suggestion by Foster that he should study a new drug, *jaborandi* (pilocarpine). Langley's main contribution from this work was histological, showing that the cells were depleted of granules during secretion. He also described the nerve supply to the glands and its effects. Over twenty papers on secretion were published between 1875 and 1890. From this work he contributed the chapter on 'Salivary Glands' to Schäfer's *Textbook of Physiology* in 1898 (see p. 149).

By 1890, Langley had begun his second phase, following on the work of Gaskell on the visceral nervous system. During the next thirty years, by careful anatomical and histological descriptions and experimental observations often using nicotine, Langley produced the now accepted description of the automatic nervous system – parasympathetic, sympathetic; preganglionic, postganglionic; white rami, grey rami are words which were given their meaning by Langley in these classical experiments. He published about forty papers on the autonomic nervous system and this work was summarised in the Croonian Lecture to the Royal Society in 1906 and a book, *The Autonomic Nervous System, Part I* published in 1921.

He also published papers on many other topics. The obituary notice in the *Journal of Physiology* (Volume 61) lists 171 publications in the fifty years from 1875 to 1925. Co-authors included Sherrington, Fletcher, Anderson and many others less known to physiologists. He worked very systematically, carrying out the series of experiments on a topic, then withdrawing from the laboratory to write the paper. Then he returned to work on the next topic.

For about forty years, Gaskell and Langley were both working in Cambridge on related fields but their approach and outlook were quite different. Henry Dale, who was a student about 1894 described his experience of the two men:

Langley had no gift of inspiration as a lecturer, but I probably got more from Langley than I recognised at the time. He was an imperturbably accurate observer, eager for an accurate account of observed facts and impatient of speculative theory. I wrote weekly essays for him, and his dry criticism, contemptuous of theory which was not merely a convenient summary of facts, was a hard, but I believe, a wholesome discipline.

I also felt a real glow of inspiration from W. H. Gaskell's lectures. Gaskell was certainly the most stimulating teacher of advanced students that I ever encountered. He gave the hearers an inspiring vision of research as an exciting intellectual adventure.

Gaskell's attitude and Langley's were in strong contrast and direct antipathy. I believe, in retrospect, that the opportunity of daily contact with men of such contrasted attitudes, each of them in the highest rank as an exponent and practitioner of his own conception of scientific research, had an educative value much greater than we recognized.

At the time I was much more conscious of my debt to the enthusiastic and unselfish interest shown in my studies and my callow ideas by H. K. Anderson, then a demonstrator in Physiology and later Master of Caius College.

As an undergraduate Langley won many prizes for short-distance running, and always fostered athletics in Trinity and the University. As a young man he rowed all over the fens of East Anglia. He played tennis and cycled all his life, particularly after the introduction of the safety bicycle in 1894. But his special hobby became skating in the English style and he contributed to the drawing up of the rules of the sport by the National Skating Association. He often took a skating winter holiday in Switzerland and if the fens became frozen he took every free hour to skate elaborate figures with a grace which many generations of undergratuates sought to emulate.

In 1902, he married and left his rooms in Trinity to live in a house on the outskirts of Cambridge towards Madingley. His garden now became his hobby; his tulips in May and roses in June were unsurpassed in Cambridge, where good gardeners abound. Here he entertained his friends. The founders of physiology in Cambridge – Foster, Gaskell and Langley – all became gardeners in their later years.

In his early graduate days, Langley belonged to discussion and dining clubs with companions interested in fields remote from physiology. Among his contemporaries were such as Hallam Tennyson, Edmund Gosse, A. C. Benson, the Lyttletons. In his rooms in Whewell's Court he enjoyed giving genial hospitality; he was not so genial to anyone who interrupted him as he sat working at night at his immense roll-top desk with a pipe in his mouth. In private life, and at committee meetings, he allowed his resentment to show towards those who wasted time. This was apparently his strongest characteristic; Fletcher says: 'It is hard to believe that in his long and strenuous life he ever spent an idle or wasted day'.

Arthur Sheridan Lea, FRS (1853–1915)

Lea was born in New York, but his family returned to England and settled in Liverpool, where he went to school. In 1872 he entered Trinity College, Cambridge and took a first class in National Sciences Tripos in 1875, MA in 1879 and DSc in 1886.

Soon after his success in the Natural Sciences Tripos, he accepted a suggestion from Michael Foster that he should devote himself to teaching and research in physiology and he immediately became Demonstrator in Foster's classes. In 1881 he was appointed Lecturer in Physiology and Assistant Tutor at Caius College, in 1885 he was elected a Fellow of that College and for two to three years acted as Bursar. In 1884 he became University Lecturer in Physiology.

His special interest was Physiological Chemistry, although he had no special training as a chemist. His contribution to the teaching was to establish an advanced course of Physiological Chemistry from 1876 until it rather lapsed with his ill-health in 1893. To continue the course Foster then persuaded F. Gowland Hopkins to join his staff in Cambridge. In Foster's *Textbook of Physiology* of 1877, Lea wrote an appendix on the chemical aspects of the subject, which later was enlarged to a separate volume, *The Chemical Basis of the Animal Body* (1892). He was elected a member of the Physiological Society in 1877, at its first annual meeting.

Lea liked the methods of research. He and Kühne, in Heidelburg in 1878, preceded Langley's description of the histological changes in gland cells during secretion by describing changes in the pancreas. In Cambridge he investigated the action of various ferments, using the method of allowing the fermentation to occur in dialysis tubes, so that the digestion products were continually removed. He was elected FRS in 1890.

But his life was a struggle against progressive spinal disease and after 1892 he was unable to work effectively despite a fine touch of fortitude. He left Cambridge in 1899 and lived at Sidcup until his death in 1915.

Albert George Dew-Smith (1848–1903)

Albert George was born at Salisbury, the son of Charles Dew; on succeeding in 1870 to some property he assumed the name of Dew-Smith. Albert George Dew went to school at Harrow and in 1868 was admitted as a pensioner of Trinity College, Cambridge, passing third class in the Natural Sciences Tripos of 1872.

As a student Dew-Smith met and became a friend of Michael Foster and worked with him in research on electrical stimulation of the frog's heart. Although he never became a working physiologist, Dew-Smith was an important support for Foster in the foundation of the *Journal of Physiology* and in the equipping of a working laboratory, as stated by Sharpey-Schafter (1927):

The establishment of a journal cannot be effected without pecuniary sacrifices – a difficulty which was largely met by a friend and pupil of Foster – A. G. Dew-Smith familiarly known to his friends as 'Dew'. Dew-Smith was a man of considerable fortune, of high culture and singularly good taste. He occupied until his marriage the rooms in Neville's Court which appertained to Foster as a Fellow of Trinity. They were charmingly furnished by the occupant, the walls being adorned with examples of Rossetti, Burne-Jones and other favourite artists of that day.

Another important service rendered by Dew-Smith to Physiology, and indeed to Experimental Science in general, was the establishment of the Cambridge Scientific Instrument Company, which for many years was carried on almost entirely by him, later by Sir Horace Darwin.

The link between the *Journal of Physiology* and the Cambridge Scientific Instrument Co. is shown by the fact that from 1880 until 1894, the *Journal* was published from the offices of the Cambridge Scientific Instrument Co. and then the Cambridge Engraving Co. in St Tibbs Row.

Through his connection with Foster, Dew-Smith was a founder member of the Physiological Society in 1876, until he resigned in 1899.

Dew-Smith was also a collector of books and of jewels and a noted amateur photographer. After his marriage he lived at Chesterton Hall, Cambridge and is buried in Histon Road Cemetery.

In Cambridge Dew-Smith met and became a friend of Sidney Colvin (1845–1927), Slade Professor of Fine Art and Director of the Fitzwilliam Museum (1873–85) and later Keeper of Prints and Drawings at the British Museum; as a Fellow of Trinity, Colvin was living in college at the same time as Dew-Smith. Sidney Colvin about 1870 met and began a close friendship with R.L. Stevenson (1850–94); apparently through Colvin, Stevenson met and became friendly with Dew-Smith. Subsequently Colvin assumed responsibility for the publication of Stevenson's works after 1890, when Stevenson was living in Samoa. Colvin also edited Stevenson's letters published in four volumes in 1899 and again in 1911. Volume II of the 1911 edition has as frontispiece a photograph of Stevenson by Dew-Smith taken about 1885. It also contains a letter in verse written by Stevenson to Dew-Smith in 1880; a note by Colvin explains: 'This letter is addressed to the late Mr A.G. Dew-Smith, who had sent him a present of a box of cigarettes. Mr Dew-Smith, a man of artistic tastes and mechanical genius, with a silken, somewhat foreign, urbanity of bearing, was the original, so far as concerns manner and way of speech, of Attwater in "Ebb Tide".'

The first G.H. Lewes Student (p. 116)

On October 10th, 1879 George Eliot wrote to Charles, son of G.H. Lewes:

I have had a delightful bit of news from Dr Foster this morning. He had mentioned to me before, that there was an Edinburgh student, whom he had in his mind as the right one to elect. This morning he writes – 'The trustees meet tomorrow to receive my nomination. I have chosen Dr Charles Roy, an Edinburgh man, and a Scotchman – not one of my pupils. He is, I think, the most promising – by far the most promising – of our young physiologists, putting aside those who do not need the pecuniary assistance of the studentship. And the help comes to him just when it is most needed: he is in the full swing of work, and was casting about for some means of supporting himself which would least interfere with his work, when I called his attention to the studentship. I feel myself very gratified that I can, at the very outset, recommend just the man, as it appears to me, for the post.' This is a thing your father would have chosen as a result of his life.

Charles Smart Roy, FRS (1854–97)

Roy was born in Arbroath, of which town Sharpey was also a native. Roy was educated in Arbroath and St Andrews, before studying medicine at the University of Edinburgh, where he graduated MB in 1875. He proceeded to MD, with a gold medal, in 1878 by a thesis 'On the influences which modify the work of the heart'. He used work done in Berlin which was published in the *Journal of Physiology* (Volume 1, 1879).

M'Kendrick (p. 202), who in 1872 was conducting the practical classes in physiology in Edinburgh, later wrote:

I was attracted by the earnestness of a fine healthy, rosy-cheeked lad who sat at the end of one of the benches; and one day, as we were working at muscular fibre, I discovered that this lad was familiar with its structure, and had a large collection of preparations of the muscular fibres of various animals prepared by his own hand. He invited me to pay a visit to his humble lodging, and there I found that he possessed a large collection of microscopic preparations, and that he was, in fact, an accomplished histologist.

Further, as showing how early his mind was directed to the special methods of research which he afterwards used to such good purpose, he had conceived a new plethysmograph, and had constructed a rough model.

Immediately after graduation Roy was resident physician at the Edinburgh Royal Infirmary under Balfour, an authority on valvular lesions of the heart. Then he went to the Brown Institute in London and was engaged in the investigation of contagious pleuro-pneumonia of cattle, until the Turko-Serbian war broke out in 1876; he became a surgeon-major in the Turkish Army in charge of a hospital at Yanina in Epirus where, in leisure periods, he designed his frog cardiometer. He then returned to continue his work at the Brown Institute.

In 1877–79 he was working in European Laboratories – in Berlin with du Bois Reymond, Virchow and Knonecker; in Strassburg with Goltz; in Leipzig with Cohnheim. In this work with German colleagues, most of the apparatus was designed by Roy, investigating the physiology of the heart with his frog cardiometer, the pulse wave with a small plethysmograph to measure the volume of the artery and an instrument to measure elasticity of the wall. He worked with Graham Brown on capillary blood pressure and then on renal blood flow using a renal oncometer. Most of this work was published in the new *Journal of Physiology* after Roy had returned to this country. Always the work derived from Roy's marvellous skill in designing new apparatus and in the operative skill with which he applied it.

Roy stayed with Cohnheim in Leipzig for nearly a year until he was appointed G. H. Lewes Student, when he went to work in Foster's laboratory in Cambridge, which in 1880 was well established. His chief work as G. H. Lewes student was on the physiology and pathology of the spleen using an oncometer he designed for the purpose; the work was published in *J. Physiol.* in 1882. Roy became a member of the Physiological Society in 1880 and in the next three years gave seven demonstrations to the Society, mostly of apparatus for experiments on the heart; later he presided at a Cambridge meeting of 1890.

In 1882, Greenfield was appointed Professor of General Pathology in Edinburgh and Roy succeeded him as Professor–Superintendent of the Brown Institute, there continuing his work on pathology and on the physiology of the heart, until in 1884 a Chair in Pathology was created in Cambridge and Roy was appointed to it. In the same year he was elected FRS. In Cambridge he was in residence in Whewell's Court, Trinity College until 1887, when he married Violet, the daughter of Sir George Paget.

Thus it fell to Roy to develop pathology as a separate subject in Cambridge. He had, perhaps, insufficient interest in anatomical and histological descriptive pathology to be a good teacher of undergraduates and it was said that he had no interest in the ordinary examination student. His interest was to create a research institute in pathological physiology and its clinical applications, and in this he was encouraged by Clifford Allbutt, then Regius Professor of Physic in succession to Sir George Paget. To young research workers Roy was superb in his friendly help and his ingenuity and patience in overcoming their difficulties. At first pathology was carried on within the Department of Physiology until in 1889 space became available for a separate Laboratory of Pathology.

Experimental work in pathology was greatly assisted by the foundation, in 1887, of the J. Lucas Walker Studentships, the students being largely selected by Roy himself. Research was carried forward on topics such as intracardial pressure, the relation between the heartbeat and the pulse wave, the specific gravity of blood, infection and phagocytosis, the formation of haemoglobin. Of outstanding importance was the work on the dog's heart published in 1890 and 1892 in conjunction with Adami, in which the contraction of various parts of the heart muscle, cardiac volume and intracardial pressures were simultaneously recorded, with specially designed apparatus. This work, like that of Newell Martin (p. 174), must be regarded as providing the initial basis from which later, more accurate, work developed, but like the work of Martin, is unknown to modern physiologists.

J. G. Adami, FRS (1862–1926) became a member of the Physiological Society in 1887. He was Professor of Pathology in Montreal from 1892 to 1919 and then Vice-Chancellor of the University of Liverpool.

After 1880 bacteriology was becoming increasingly important. The rather monotonous nature of its laboratory work did not appeal to Roy but he appreciated the fields which were opening. At the Brown Institute he had been involved with infective disease in cattle and visited Argentina to investigate and devise an inoculation against a disease amongst the cattle herds. In 1885, soon after he had become professor at Cambridge, he led a small commission sent to Spain to investigate a severe form of cholera. His companions were Sherrington, then just at the end of his student days and J. Graham Brown (1853–1925) who was an Edinburgh physician with whom Roy had worked on the capillary circulation; he was the father of T. Graham Brown, FRS (1882–1965), a student of Sherrington, Professor of Physiology in Cardiff (1920–47) and a famous Alpine climber.

Roy's extraordinary career was cut short by nervous illness which prevented him from working after 1894 and led to his death in 1897, aged forty-three. After a service in the chapel of Trinity College he was buried in Mill Road Cemetery, Cambridge.

References

Breathnach, C. S. (1969). 'Henry Newell Martin (1848–1893). A pioneer physiologist', *Medical History*, **13**, 271–9 (P).

Dale, H. H., cited by Feldberg (1970) in *Biographical Memoirs of Fellows of the Royal Society*, **16**, 88–9.

Fye, W. B. (1985). 'H. Newell Martin – a remarkable career destroyed by neurasthenia and alcoholism', *Journal of the History of Medicine*, **40**, 133–66.

Sewall, H. (1911). 'Henry Newell Martin, Professor of Biology in Johns Hopkins University, 1876–1893', *Johns Hopkins Hospital Bulletin*, **22**, 327–33.

Sharpey-Schafer, E. (1927). *History of the Physiological Society during its first Fifty Years, 1876–1926*. Issued by the Society and published as a supplement to *Journal of Physiology*, December. Cambridge University Press.

Biographies

Adami: *British Medical Journal* (1926), ii, 507–10 (P).
Babington: *Proc. Roy. Soc. London* (1895–6), **59**, viii–x.
Darwin, H.: *Proc. Roy. Soc. London A* (1929), **122**, xv–xviii (P).
Foster: *Proc. Roy. Soc. London B* (1907–8), **80**, lxxi–lxxxi (P) (by Gaskell).
J. Physiol. (1906–7), **35**, 233–46 (P) (by Langley).
British Medical Journal (1907), i, 349–51 (P).
Gaskell: *Proc. Roy. Soc. London B* (1915), **88**, xxvii–xxxvi (by Langley).
Lancet (1914), ii, 869–70 (P).
Humphry: *Proc. Roy. Soc. London* (1905), **75**, 128–30.
British Medical Journal (1896), ii, 975–80 (P).
Langley: *Proc. Roy. Soc. London B* (1927), **101**, xxxiii–xli (P) (by Fletcher).
J. Physiol. (1926), **61**, 1–27 (P) (by Fletcher).
British Medical Journal (1925), ii, 923–5.
Lea: *Proc. Roy. Soc. London B* (1915–17), **89**, xxv–xxvii (by Langley).
Martin: *Proc. Roy. Soc. London* (1896), **60**, xx–xxiii (by Foster).
Sewall (1911).
Breathnach (1969) (P).
Fye (1985).
Newton: *Proc. Roy. Soc. London B* (1907–8), **80**, xlv–xlix.
Paget, G.: *Proc. Roy. Soc. London* (1891–2), **50**, xiii.
Lancet, 1892, i, 392–4.
Roy: *Proc. Roy. Soc. London* (1905), **75**, 131–6 (by Sherrington).
British Medical Journal (1897), ii, 1031, 1124.

12

Physiologists at King's College, London

The developing department

By succession of Todd, Bowman, Simon and Beale (see Ch. 4), King's College in 1868 had become a centre of excellence in histology. In 1869, Beale (p. 43) resigned his post as Professor of Physiology, wishing to devote his time to clinical work and possibly also to expressing his ideas in relation to Darwinism. Hearnshaw (1929) also suggests that he appreciated that it was now necessary for a physiologist to use experimental methods of which he had no experience and did not wish to take the time necessary to become master of these new methods.

The appointment of William Rutherford of Edinburgh as Professor of Physiology fulfilled the need that the college should have a professor who could use the new methods of physiology; Rutherford had been assistant to Hughes Bennett in Edinburgh (p. 96).

The arrival of Rutherford forced great development of physiology in King's. He instituted practical classes of the type he had helped set up in Edinburgh, with the students preparing their own microscopic sections, and carrying out experiments in chemical physiology. In Rutherford's ideas experimental physiology was best shown by elaborate demonstrations attached to lectures and an example of such a course of lectures was published by the *Lancet* in 1871 (see p. 132). By 1871 the activity in physiology was such that a demonstrator was required and Ferrier, who became a great experimental physiologist, although his main work was clinical, was appointed. Growing activity required more laboratory space and letters from Rutherford to the Council of King's became chronic, until a new site became available at the south-east corner of the building, when the embankment of the Thames was made. The plan was determined by 1872 but there were delays which meant that the laboratories could not be opened until 1875, after Rutherford had been appointed professor at Edinburgh; it was therefore Yeo, as professor, and Ferrier who put this accommodation into full use. It was used by physiology until 1900 when extensive accommodation for medical and scientific departments was created by extensive rebuilding of the whole of King's College.

When Rutherford resigned in 1874 he was succeeded by Yeo. Yeo's assistant in the classes was Ferrier, who at that time was Professor of Forensic Medicine.

Until Rutherford, the accepted practice at King's had been that the Professor of Physiology should also be a physician at the hospital and this difficulty had to be circumvented to allow Beale to continue his hospital work; he was given the title of Professor of Pathological Anatomy. Rutherford had no wish for clinical responsibility and held no post in the hospital. At first Yeo was also assistant surgeon, until he resigned from clinical work in 1880. Although both Yeo and Ferrier were part-time physiologists they were also active experimenters, continuing the work on the brain which Ferrier had begun in 1873 at the West Riding Asylum (p. 192). While he was at King's Rutherford also continued with original experimental work.

Thus over the period 1869–80, by the efforts of Rutherford, Ferrier and Yeo, experimental physiology was established at King's College, London. These men were important contributors to the growing status of experimental physiology in the years after 1880.

Professors

William Rutherford (p. 198)

William Rutherford was professor at King's for only five years, 1869–74; since the rest of his life was in Edinburgh the full life of Rutherford is given in the next chapter.

However his short period at King's was critical to the development of physiology in the College as has been described above. Firstly he introduced experimental physiology to the teaching of students, adding to already existing skills in the teaching of histology. Rutherford was highly competent to introduce this new method of teaching by his experience in Edinburgh and by his experience while in London of demonstrating in Huxley's classes at South Kensington (p. 113). Rutherford was a man forceful in getting what he wanted and so created the laboratories at King's. These were also available for research work.

Secondly, Rutherford decided that he would do no clinical work, and so for the first time at King's the professor was a full-time physiologist, not sharing his attention with clinical work. Before 1870 the only full-time physiologist had been Sharpey at University College and he was now joined by the professor at King's. It was many years before other London Schools had full-time lecturers or professors (Ch. 14).

Rutherford's years in London were probably the best in his life. He was free to develop his own methods of teaching, including the elaborately illustrated lectures and demonstrations. He did some useful research work without interference and was in full health. Accepted by his peers, Sharpey, Huxley and Foster, Rutherford was immediately invited to be an original member of the Physiological Society and soon after, in 1878, was chosen by Foster as a co-editor of the *Journal of Physiology*. These years contrast wtih the difficulties which later arose in Edinburgh.

Gerald Francis Yeo, FRS (1845–1909)

Gerald Yeo was born in Dublin, went to school in Tyrone and received his medical education at Trinity College, Dublin, graduating MB in 1867 and MD in 1871, with a gold medal for an essay on diseases of the kidneys. He entered practice in Dublin, taught in anatomy and also pursued further studies in Paris, Berlin and Vienna; he was fluent in European languages.

In 1874, Yeo was appointed Professor of Physiology and Anatomy at King's College, London and was also appointed Assistant Surgeon at King's College Hospital. He resigned the clinical appointment in 1880 to devote his full time to physiology. Yeo was elected FRS in 1889 and remained professor at King's until 1890, when he retired into private life. He dearly loved country pursuits and had a country home at Fowey where he was able to share with physiological colleagues his love of sailing and the sea. After he retired he lived in Totnes and only rarely, usually for scientific meetings, went to London. His biographers recall that he was a delightful companion.

As a teacher he continued the practical classes which Rutherford had started and indeed inherited the laboratory which Rutherford had fought to obtain. He seems to have been an effective lecturer and wrote a *Manual of Physiology* which was a popular student book, requiring six editions during his years at King's. From King's he published experimental work on bile, heart muscle, skeletal muscle and heart sounds. The famous experimental work from King's about 1880 was the work on cerebral localisation in the monkey which Yeo carried out in collaboration with Ferrier. The work was essentially the work of Ferrier, but Ferrier had not been allowed a vivisection licence under the 1877 Act and Yeo did the operative work, a fact which did not emerge until Ferrier was prosecuted under the Act (p. 193).

Yeo contributed actively to the publicity campaign against anti-vivisectionists, supporting the need for experiments in physiology. His Arris and Gale lectures in 1882 to the College of Surgeons consisted of a forceful and often amusing account of how little had been achieved in 2,000 years without experiments and how much had been achieved by experiments in the last thirty years.

Yeo was the first secretary of the Physiological Society. He was one of the nineteen men who met in Burdon Sanderson's house and resolved to form the society and as soon as its constitution was adopted, Yeo and Romanes were chosen as secretaries (p. 262). Yeo kept the minutes, but kept them indifferent well (Sharpey-Schafer, 1927); he remained secretary until he retired in 1890, when the society showed its gratitude by presenting him with a silver tea service. His nature and linguistic skill allowed him to become the effective organiser with Foster of the first International Congress of Physiology held in Basle in 1891. The idea had originated in a meeting of a group of friends, including Kronecker of Berlin and Michael Foster, during the International Medical Congress in London in 1881; Yeo and C. S. Roy (p. 183) were joint secretaries of the Physiological Section of the 1881 Congress in London.

Research workers

Sir David Ferrier, FRS (1843–1928)

David Ferrier was born in Aberdeen and attended grammar school there, being placed first in the competition for bursaries and thus entered the Faculty of Arts at Aberdeen University. He graduated MA in 1863 with first class honours in classics and philosophy and then for a short period studied psychology in Heidelberg. His early work in philosophy possibly influenced his later choice of interest. He was a pupil at Aberdeen of Alexander Bain, author of a well known treatise 'Body and Mind' and who certainly thought of the 'spirit of man' in relation to the anatomical nature of the brain. In 1865 Ferrier returned to Edinburgh, entered the Medical School and graduated with all possible honours in 1868.

He was still uncertain what he wanted to do. For a short time he was assistant to the Professor of Practical Medicine, he tried coaching students because he needed money, and then became assistant to a general practitioner at Bury St Edmunds in Suffolk. Ferrier himself said he was not a success as a general practitioner but Dr Image was an accomplished man of a family well known for scholarship and proficiency in natural science. With the active co-operation of Dr Image, Ferrier spent his time in research, an investigation into 'the Comparative Anatomy and intimate structure of the Corpora Quadrigemina'; this was carried out in Dr Image's garden, locally famous for its beauty. The work was sent in as an MD thesis to the University of Edinburgh and was awarded a gold medal.

In 1870 Ferrier went to London, probably encouraged by Burdon Sanderson, whom he succeeded as Lecturer in Physiology at the Middlesex Medical School. After only one year, Ferrier moved to King's College, London as Demonstrator in Practical Physiology and remained connected with King's for the rest of his working life. The Professor was Rutherford and, in addition to the classes at King's, Ferrier was also one of the demonstrators in Huxley's classes at South Kensington (p. 113). In 1872 he gave up his appointment in physiology to become lecturer and then Professor of Medical Jurisprudence and was also appointed assistant physician at King's College Hospital and at the West London Hospital. While lecturing on Medical Jurisprudence he was co-author of the later editions of Professor Guy's book on forensic medicine.

From 1872 onwards Ferrier was carrying out his classical experimental work on the localisation of function in the cerebral cortex (see below). The importance of this work was immediately recognised; he was chosen to give the Croonian Lecture in 1874 and 1875 and was elected Fellow of the Royal Society in 1876. His work was also published in a book, *The Functions of the Brain* (1876). He combined his experimental work with clinical work in King's College and West London Hospitals and, after 1880, as physician to the National Hospital for the Paralysed and Epileptic, Queen Square, where he joined a famous staff including Hughlings Jackson. He also had a large private practice. His clinical observations, parallel with his physiological experimentation, led to an invitation to deliver the Goulstonian Lectures to the College of Physicians in 1878, under the title of 'The Localisation of Cerebral Disease'. In 1889, King's College recognised his eminence by creating for him a personal Professorship of Neuropathology and at the same time he became Physician to the Hospital. He retained these posts until 1908, when he became Consulting Physician at King's.

Ferrier achieved great public honour as a clinician, being particularly recognised by lectureships and awards in the College of Physicians, and was knighted in 1911. He was always regarded as a man who, by research work and his meticulous consideration of the results of lesions, was advancing accurate knowledge of neurology. On the other hand the traditions of the National Hospital, Queen Square say that, although he had a large private practice, he lacked the faculty of painstaking clinical examination and always gave the impression of being primarily a physiologist.

Physiologists then and afterwards have accepted that Ferrier's work on the localisation of function in the cerebral cortex began modern neurophysiology, but Ferrier was a professional physiologist only for the years 1870–3, when he taught physiology at the Middlesex and King's. His experimental work was done under a series of peculiar conditions. He set out to confirm and develop existing work which indicated that movements could be produced by stimulation of the cortex. In 1873 he went to visit his friend J. Crichton-Browne (see below), who was then superintendent of the West Riding Lunatic Asylum at Wakefield and had established a laboratory there. In this laboratory Ferrier carried out experiments stimulating the cortex of pigeons, fowls, guinea pigs, rabbits, cats and dogs provided by the laboratory. The results were immediately published in the West Riding Lunatic Asylum Reports of 1873; from these reports Crichton-Browne and Ferrier later formed the journal *Brain*. In London, Ferrier at first kept his animals in sheds at the back of his house until, after the Act of 1877, they were kept at King's College. Ferrier's work was now extended to monkeys under a grant from the Royal Society but Ferrier himself was refused a licence under the Act. The operative work was done by Yeo who had the necessary licence for this work. The work was thus done in the Physiology Laboratory of King's College which Rutherford had fought to establish.

Ferrier was a founder member of the Physiological Society; after the initial meeting at Burdon Sanderson's house, he was one of the small group who drew up the constitution of the Society. At the Jubilee Meeting in 1926, he was elected an Honorary Member. The high opinion of him amongst physiologists was shown at the opening of the Physiological Laboratory in Cambridge in 1914, when the University of Cambridge conferred on David Ferrier the honorary degree of Doctor of Science. He was always associated with the *Journal of Physiology* and when the *Journal* was taken over by

Langley in 1894, Ferrier was listed as one of the Advisory Committee and he continued in this capacity until his death.

Ferrier's work attracted the attention of the antivivisectionists, particularly after the International Medical Congress in London in 1881. In the Section of Physiology, a discussion on Localisation of Functions in the Cerebral Cortex was opened by Goltz, who presented his evidence that extensive destruction of the cortex in dogs produced no specific disabilities; the second speaker was Ferrier who described the precise effects produced by localised lesions in monkeys. Goltz had brought from Strasbourg one of his dogs and this animal, with two of Ferrier's monkeys, was shown next day to members of the congress in the laboratory at King's College. The animals were then killed, the brains shown to the members and then handed to a committee of Klein, Langley, Purser and Schäfer for detailed anatomical and histological report. By this episode, the idea of localisation of function in the brain was firmly established and Ferrier was acclaimed as an initiator in physiology.

But the episode had attracted the attention of the antivivisectionists, who promoted a prosecution of Ferrier, accusing him of experimenting without a licence under the 1877 Act — indeed he had been refused one. Legally the case was soon dismissed because the operative work had been done by Yeo, not Ferrier. The case became propaganda for both sides: the antivivisecionists claimed that it showed the Act to be unable to prevent gross cruelty; physiologists claimed that it showed that the Act was severely hindering proper and necessary research work. For Ferrier himself, the publicity increased the number of private patients who consulted him.

The biography of Crichton-Browne added below is perhaps out of place: he was never attached to King's College.

Sir James Crichton-Browne, FRS (1840–1938)
Crichton-Browne was not a physiologist but was the pioneer in England of the study and care of the insane. For a short period from 1868–75, when he was its director, the West Riding Asylum at Wakefield was a centre of neurological interest, where many eminent men congregated and made their contributions at monthly *conversaziones* organised by Crichton-Browne. Among them were Hughlings Jackson, Ferrier and Allbutt.

His father was W. A. F. Browne, the first medical superintendent of the Crichton Royal Mental Hospital at Dumfries. He was christened

James Crichton Browne (and later adopted the hyphenated name) in tribute to Crichton who had founded the institution at Dumfries. James qualified MD at Edinburgh in 1862 and immediately took up psychiatric medicine, serving as assistant in asylums in Devon, Derby, Warwick and Newcastle upon Tyne before being appointed Medical Superintendent of the West Riding Asylum at Wakefield in 1866.

It was here that he initiated research into mental disease and decided that the most likely approach was by examination of the brains of patients who had suffered various forms of insanity. He therefore established a laboratory and in this laboratory David Ferrier carried out some of his early research. Since facilities for publication were poor, Crichton-Browne in 1871 initiated the West Riding Asylum reports in which the work of Ferrier was published in 1873 and in which papers by Jackson, Allbutt and others also appeared.

Crichton-Browne always advocated the early treatment of mental disease and thought that the students should be taught about it. He offered to give six lectures on mental diseases at the Medical School in Leeds, and a weekly clinic at the asylum, with provision of luncheon, provided the students were select and not too numerous. In fact the class had an attendance of eighteen, the largest class of psychological medicine in the country.

In 1875 he was appointed Lord Chancellor's Visitor in Lunacy. The Asylum at Wakefield continued to be an important centre for research and treatment, but Crichton-Browne himself had now little opportunity for original work. He remained Visitor until he retired in 1922. With Hughlings Jackson and Ferrier in 1878 he founded the journal *Brain* to succeed the West Riding Asylum Reports and was one of its editors until it was taken over in 1888 by the Neurological Society, which he had largely formed. He was knighted in 1886 and elected fellow of the Royal Society in 1883. He was given the honorary degree of Doctor of Science by the University of Leeds in 1909.

Throughout his long life he continually advocated the study of mental diseases and their early recognition and management. He was a gifted orator who gave many addresses on this subject and in the later part of his life was a famous after-dinner speaker. He wrote several volumes of reminiscences under the title of *Leaves from a Doctor's Diary*, the last of which appeared at about the time of his death at Dumfries in 1938, aged ninety-seven years. He had no connection with the Physiological Society.

Comparative anatomists

Alfred Henry Garrod, FRS (1846–79)

Alfred Henry Garrod was born in London. He was the son of Sir Alfred Baring Garrod, FRS (1819–1907), physician at King's College Hospital and a great authority on gout. A younger son, Archibald Edward Garrod (1867–1936) was also elected FRS and became Regius Professor of Medicine at Oxford. Thus a father and two sons were all fellows of the Royal Society.

Alfred Henry attended University College School until he entered as a science student at University College, London, where he was much influenced by Sharpey. For medical education he went in 1864 to King's College, London with a Warneford Scholarship and was there a good student, winning the medical scholarship in three consecutive years; he qualified Licentiate of the Society of Apothecaries in 1868.

In the meantime, he had matriculated in the University of Cambridge, firstly at Gonville and Caius College in 1867, migrating in 1868 to St John's of which he was elected scholar in 1870. Garrod passed first class in the Natural Sciences Tripos in 1871, and graduated BA in 1872 and MA in 1875. In 1873 he was elected to a Fellowship at St John's, the first student of natural sciences to be so elected in that college.

Garrod was appointed prosector to the Zoological Society in 1871 and thereafter the dissecting room of the Society in Regent's Park was his work place. He became an authority on the anatomy of birds but also dissected any animal which became available – it is noted that he dissected five rhinoceroses.

While holding this post at the Zoo, Garrod was also appointed Professor of Comparative Anatomy at King's College, London (1874–79) and Fullerian Professor at the Royal Institution (1875–79). He was elected FRS in 1876. His lectures at the Royal Institution on 'The Classification of Vertebrate Animals' (1875), 'The Human Form' (1877) and 'The Protoplasmic Theory of Life' (1878) were very popular and reveal the wide range of his interests. There were many papers published in *Proceedings of the Royal Society, Journal of Anatomy and Physiology, Nature*, etc. and he was for several years a subeditor of *Nature*. He died of tuberculosis in 1879, aged thirty-three.

At the foundation of the Physiological Society in 1876, Garrod declined an invitation to become a founder member, presumably because he then regarded himself as a comparative anatomist. Earlier, in 1869−73, he had published some ten papers in *Journal of Physiology and Anatomy* on physiological topics, such as the cause of diastole, cardiographic recording through the chest, the sphygmograph, body temperature and environment, factors determining heart rate.

Francis Jeffrey Bell (1856−1924)

Jeffrey Bell was the son of F. J. Bell of Calcutta and a relation of Charles Bell (p. 3). He entered Magdalen College, Oxford in 1874, studying under Rolleston, and took second class in Natural Sciences in 1877. He became BA in 1878 and MA in 1880.

Immediately after graduating he was appointed assistant in the Zoological Department of the British Museum and stayed in this post until his retirement in 1919. Also from 1879−96 he was Professor of Comparative Anatomy at King's College, London in succession to Garrod. In 1885 Bell published a *Manual of Comparative Anatomy and Physiology* which was much used by medical students.

Bell's main publications concerned his work at the British Museum, where he had charge of echinoderms and worms and maintained good displays in his exhibition gallery. His most important publication was *A Catalogue of British Echinoderms in the British Museum* (1892) and he also wrote reports on the echinoderms of the 'Alert', 'Southern Cross' and 'Discovery' expeditions.

He was elected member of the Physiological Society in 1880 but made no contributions to its proceedings, although he was a regular attender at meetings.

Botanist

James William Groves

According to Sharpey-Schafer (1927), at the time of his election to the Physiological Society in 1880 Groves was assistant to Yeo at King's College, London particularly teaching histology. He had been secretary of the Medical Microscopic Society whose functions were largely replaced by the formation of the Physiological Society.

Later, from 1887−92, Groves was Professor of Botany at King's College and was apparently a somewhat strange figure. Hearnshaw (1929) writes: 'He kept his chair for scarcely six years. At the end of 1892 he resigned, in order to travel to regions where vegetation flourished more luxuriantly than in the Strand. He broke the shock of his departure by presenting to the Council "a life-sized female figure in plaster entitled *The First Plunge*", which either difficulties of transport or natural modesty prevented him from taking with him to the continent.'

Groves made two demonstrations to meetings of the Physiological Society at King's in 1882, 1883 on histological techniques. He was a regular attender at meetings.

References

Hearnshaw, F.J.C. (1929). *The Centenary History of King's College, London*. London: Harrap.

Biographies

Bell: *Nature* (1924), **113**, 541.
Crichton-Browne: *Lancet* (1938), i, 406−7 (P).
Ferrier: *Lancet* (1928), i, 627−9 (P).
Proc. Roy. Soc. London B (1928), **103**, viii−xvi (by Sherrington).
Garrod: *Dictionary of National Biography*, **21**, 27−8 (by Bettany).
Groves: Hearnshaw (1929).
Yeo: *Nature* (1909), **80**, 314−15.
British Medical Journal (1909), i, 1158.

13

Physiologists in Scotland

Laboratories in 1871

In 1870 none of the Scottish universities had a Professor of Physiology. In the next decade Rutherford in Edinburgh, M'Kendrick in Glasgow and Stirling in Aberdeen were appointed Professor of the Institutes of Medicine and, by taking no clinical appointments, turned their departments into Departments of Physiology. The theme of this chapter is the replacement of the old Institutes of Medicine by Physiology, by men who were all pupils of Hughes Bennett (p. 96).

In 1871 Burdon Sanderson (Nature, 3, 189) could list four laboratories in Edinburgh where experimental work was being done but none in Glasgow or Aberdeen. In Edinburgh there was the University Laboratory of Hughes Bennett where practical physiology was being taught to students (p. 97). Burdon Sanderson also referred to laboratories of Christison (p. 100) and Maclagan (p. 101) where experimental work was being done, for example by Gamgee and Brunton. For a short period about 1870, Gamgee was Lecturer in Physiology at the Surgeon's Hall, Edinburgh and Burdon Sanderson states that he had a laboratory there. Physiology was beginning in Edinburgh in 1870 and was soon greatly developed by Rutherford. In contrast, after 1876, M'Kendrick in Glasgow and Stirling in Aberdeen had to develop Physiology as a new subject.

Edinburgh

William Rutherford, FRS (1839–99)

Rutherford was a Scot whose working life, apart from five years at King's College, London, was spent in Edinburgh. He was born at Ancrum, Roxburghshire, was educated at Jedburgh Grammar School and then went to the University of Edinburgh, where he graduated in medicine in 1863. After junior hospital appointments, he spent about a year in continental medical centres – Berlin, Vienna, Paris and Leipzig where he worked with Ludwig. In 1865 he returned to Edinburgh and was appointed assistant to Hughes Bennett, Professor of the Institutes of Medicine.

Bennett had already introduced classes in practical physiology and these were further developed by Robertson (p. 99) and Rutherford in their time as assistants. Also Rutherford carried out research work on the effect of the vagus on the circulation and made histological examinations of the brains of insane people.

When only thirty years of age, in 1869 Rutherford was appointed Professor of Physiology at King's College, London. Rutherford brought from Edinburgh experience and expertise in practical classes in physiology and so he introduced courses in practical physiology – histology, chemical and experimental – to King's, and was responsible for the establishment there of a properly equipped laboratory (p. 188). With his established interest in practical classes as a teaching medium, Rutherford became associated with Huxley as a demonstrator in the classes at South Kensington (p. 113). Rutherford also lectured as Fullerian Professor at the Royal Institute, in succession to Huxley. At King's, Rutherford could follow his own system of teaching, particularly the use of elaborate demonstrations as part of his lectures. A series of such lectures was published in the *Lancet* (1871, i and ii) and there he also explained his arrangements for teaching histology, chemical physiology and practical physiology. Foster in Cambridge and Huxley did not favour lecture-demonstrations as a mode of teaching (see p. 166). At this time it was still usual for the professor also to be a physician or surgeon in King's College Hospital. Rutherford opposed this custom and accepted no clinical post. The importance of this brief period in London has been discussed in the previous chapter.

In 1874, Bennett in Edinburgh fell ill and resigned his chair. Rutherford applied for the post and was appointed Professor of the Institutes of Medicine. The election of a successor to Bennett was used by the *British Medical Journal* to attack the system by which candidates asked for testimonials from their friends, which were incorporated into a pamphlet to be sent to the electors, but was also generally circulated. The pamphlet of Pettigrew, then Professor of Anatomy at Glasgow, was particularly attacked as an example of the evils of the system. Hughes Bennett himself wrote to the *Lancet* (1874, ii, 534) objecting to a claim made by Rutherford in his letter to the curators that he had organised in Edinburgh in 1867 two new courses in experimental physiology and practical physiological chemistry – a claim supported on obviously incomplete understanding by Sharpey and Foster. In fact the classes had been started before Rutherford became assistant (p. 98); he added to them but did not initiate them.

As at King's, on his appointment at Edinburgh, Rutherford accepted no clinical appointments. Bennett had followed the established role of Professor of the Institutes of Medicine by being also Physician to the Hospital and indeed most of Bennett's work was as a scientific physician. Rutherford, on the contrary, took no part in clinical work or teaching and devoted all his time to teaching and research in physiology. Interrupted only by a period of mental breakdown, he continued his spectacular teaching until his death in 1899. Whatever the other aspects of his life, Rutherford was in the great tradition of Edinburgh teachers of anatomy and physiology. The affection and enthusism he aroused in students was displayed at his funeral. He was buried at Ancrum, the procession being preceded by 400 students, walking four abreast.

His life was devoted to teaching physiology. He was a specialist in the preparation of tissues for microscopic examination and his methods were made available by a book *Outlines of Practical Histology*, first published in 1875. He modified existing microtomes and was one of the first to develop and use the freezing microtome.

It was his delight to illustrate his lectures by his own diagrams and live demonstrations, as is illustrated by the set of lectures published in the *Lancet* of 1872. A model of the circulation that he used is the basis of a diagram in many modern textbooks. His attitude to lecturing was expressed by a biographer in the *British Medical Journal* (1899, i, 564): 'He determined that his lectures should be illustrated as no lectures had ever been illustrated before. His love of art and untiring energy enabled him, by himself and by superintending others, to produce in an incredibly short time a set of admirable diagrams for his course, while the most precise and delicate experiments were, often with hours of work, prepared daily for the experimental table.'

The perfection of the demonstrations and the clarity of the lectures, combined with personal individualities in his manner, made him a favourite with the students, as witnessed at his funeral.

Such intense work in preparing his teaching limited his opportunities for research, although he did write on many topics without great concentration of effort on any one. He studied the effects of the vagus on the circulation and investigated histological changes in brain lesions and insanity. While at King's he worked on the spread of excitability along spinal nerves. His largest work was with Vignal after he returned to Edinburgh. They investigated the effect of drugs on the secretion of bile, involving cannulation of the bile duct. The experiments were on dogs paralysed by curare without other medication and, because a home office licence was not granted to Rutherford, they were carried out in France. Rutherford was thereafter subjected to vicious personal attacks by antivivisectionists. A sensitive and humane man, he was deeply hurt by these attacks. He was not a great experimental research worker but did make contributions in many fields and certainly played a part in establishing physiology as an experimental subject. He was elected FRS (London) in 1876.

He was a founder member of the Physiological Society, and proposed the toast of 'Our Guests' at the first dinner. In 1878, when Michael Foster founded the *Journal of Physiology*, Rutherford was named as a co-editor. For the years from 1866–74, Rutherford and others, including Ferrier, produced for each number of the *Journal of Anatomy and Physiology* some twenty pages entitled 'Progress in Physiology' consisting of reviews of physiological papers published in the six-month period.

His influence on physiology after 1890 was through a succession of assistants, who, after learning experimental physiology in Edinburgh, carried his teaching to other institutions. Of his assistants, W. Stirling became Professor of Physiology at Aberdeen (1877) and Manchester (1885), T. Anderson Stuart at Sydney (1883), de Burgh Birch at Leeds (1884), J.B. Haycraft at Birmingham (1881) and Cardiff (1894), and E.W.W. Carlier at Birmingham (1900). At the time of Rutherford's death Carlier was senior lecturer in Edinburgh and acted as deputy professor until Schäfer took over. These pupils certainly introduced quality teaching in their new appointments but in general were not great research workers.

His biographers in the *Lancet* and *British Medical Journal* were clearly former students who loved and respected him but they had some difficulty in paying their tributes to him as a person. He was an active presence in the medical social life of Edinburgh and derived great enjoyment from club life and social gatherings, particularly where there was music (see p. 102). He was a strong singer and composed songs which he himself would sing. He had a manner and mode of speech regarded by some as supercilious, haughty, vain, careless of the feelings of others. His friends said that this was a brusqueness covering innate shyness, the diffidence of an exceedingly sensitive man. He was at heart a kind man, generous with his money. His sensitivity was seriously harmed by the attacks of antivivisectionists. This is summarised in the biography in the British Medical Journal: 'Proud, shy, sensitive, resentful of injustice, he was no Laodicean, but one whose moral qualities made him loved and hated, misunderstood and admired, misjudged and enthusiastically applauded.'

Such a man inevitably became a contentious figure with many enemies in his own lifetime. Equally, Rutherford was much loved by friends, who remained loyal despite some curious behaviour in the later years of his life. After he returned to Edinburgh, he apparently lost the sympathy of his former associates in London and Cambridge. He rarely attended meetings of the Physiological Society, then held in London, Oxford or Cambridge. Despite invitations, no meeting was held in Edinburgh during his lifetime and no obituary notice appeared in the Proceedings of the Royal Society of London, although he had been elected a fellow in 1876. In his *History of the Physiological Society during its First Fifty Years, 1876–1926*, Sharpey-Schafer (1927) gives no account of the life of Rutherford in contrast to the appreciative accounts of other founder members.

Glasgow

John Gray M'Kendrick, FRS (1841–1926)

M'Kendrick, to his dislike often spelt McKendrick, was born in Aberdeen, but was almost immediately left an orphan, cared for at first by a friend and then by grandparents. He went to a village school in Perthshire and his first job at the age of thirteen was as a 'herd laddie' on a hill near Braco. The following year he was employed on a pittance as apprentice to a legal firm in Aberdeen. Before and after his seven hours in the office, he undertook an elaborate course of home study, which came to include botany and zoology, especially marine biology in the Bay of Nigg. He formed a small aquarium, which he showed at the British Association meeting in Aberdeen in 1859, attracting the attention of Huxley and other members. Professor Redfern of Aberdeen suggested that he enter medicine, and he passed the preliminary examination in the University of Aberdeen in 1861 and endeavoured to continue his studies while still employed in the law office. This was becoming impossible, but his dismissal from the lawyers' office came from his love of music; listening to Jenny Lind at a concert, he lost all sense of time and was late back to work.

He moved to Edinburgh and enrolled in medical classes, earning 12*s* per day for eight hours' work for a newspaper; however, the hours of this work let him attend the courses. He gained a gold medal in anatomy and was beginning to earn money by coaching when he suffered a pulmonary haemorrhage. After some months in the country he resumed work in Aberdeen and then in Edinburgh, supporting himself by coaching while he continued his medical studies. Finally he graduated MD, Aberdeen, in 1864.

He was appointed Medical Officer at Chester Royal Infirmary and then Resident Officer in the Eastern Dispensary, Whitechapel. Practice there was very rough but he was able to attend lectures by Huxley on comparative anatomy and by Hoffmann in chemistry. He applied for a variety of posts, including Prosectorship at the Zoo and a job on a tea plantation in Assam before, in 1866, he was appointed Resident Surgeon at the Belford Hospital, Fort William, possibly because he expressed proper religious convictions. Here he was happily settled and married when, in 1869, he met Hughes Bennett, Professor of the Institutes of Medicine in Edinburgh. Hughes Bennett had been called to Fort William in consultation and he and M'Kendrick discussed the work on calomel then being done in Edinburgh by Rutherford, Bennett's assistant. Bennett invited M'Kendrick to become his assistant when Rutherford went to King's College, London and so at the rather advanced age of twenty-nine, M'Kendrick entered physiology, adding to his income by coaching for the final examinations.

In Edinburgh, M'Kendrick was immediately involved in teaching practical physiology, both histology and experiments. At this time the Institutes of Medicine certainly had experimental apparatus but the role of Hughes Bennett is somewhat uncertain. In obituary notices of Hughes Bennett (p. 97) and in letters he wrote to the *Lancet*, it is indicated that the apparatus was introduced by Bennett; according to Noël Paton in the obituary notice of M'Kendrick in the *Proceedings of the Royal Society*, Bennett had little interest or knowledge of this apparatus and the practical class teaching was developed firstly by Robertson and Rutherford and then by M'Kendrick, officially his assistants. In fact after 1870, Bennett's failing health threw most of the teaching on to M'Kendrick but, when Bennett resigned in 1873, Rutherford was recalled from King's College. M'Kendrick did not work with Rutherford; he purchased from Bennett his diagrams, microscopes and apparatus and set up as an extra-mural lecturer. Amongst his classes was a separate class for ladies, for which there were 150 enrolments. He was also appointed lecturer at the Dick Veterinary College.

In 1876, M'Kendrick was appointed Regius Professor of Physiology in Glasgow in succession to Buchanan (p. 102). The department there had nothing except 'a few ancient microscopes and a sphygmograph' (Paton, 1926). At first both practical classes and research used the apparatus M'Kendrick had purchased from Bennett and brought with him to Glasgow. M'Kendrick also bought other equipment and it was said that in his first years he spent £1,000 on apparatus and his net income was £8. Teaching was then his forte and as his classes grew, money became available for a new building in 1903. M'Kendrick retired in 1906 leaving a new, well equipped institute to his successor, Noël Paton. At his retirement his friends showed their appreciation by a portrait and by providing money for a Laboratory of Experimental Psychology as a fitting recognition of his work on the special senses. M'Kendrick's status as a teacher led to his publishing books for students, especially his *Textbook of Physiology*, in two volumes, (1888–89).

During his career, M'Kendrick produced a considerable amount of research. His interest in music led him to investigate the appreciation of sound and in 1873 he investigated electrical phenomena in the optic nerve when light fell on the retina. About this time he visited Europe and met the continental physiologists, particularly Helmholtz, whom he revered and about whom he later wrote *A Life of Helmholtz* (1899). M'Kendrick analysed some of Helmholtz's theories by studying the curves cut on the wax cylinder of early phonographs. In addition to this work on the special senses, he worked on the relationship of nerve and muscle and, with Dewar, he studied the chemistry and physiology of chinoline and pyridine bases. This led to the discovery of various therapeutic agents and was a start towards modern pharmacology, relating function to chemical constitution. Most of this work was published in the *Proceedings of the Royal Society of Edinburgh*. With a colleague, Snodgrass, he published a book, *The Physiology of the Senses*. In 1900 with Gray, he wrote the chapters on 'The Ear' and 'Vocal Sounds' for Schäfer's textbook (p. 149).

M'Kendrick was a founder member of the Physiological Society. He was elected FRS in 1884 and served on the Council of the Royal Society, London from 1892–93.

M'Kendrick was genial, kindly and during his life developed a wide culture, beyond his specialty of physiology. He was always a popular lecturer to general audiences and in his later years wrote poetry and some philosophical books. In his retirement he lived at Maxieburn in Stonehaven where he took an active interest in local affairs and was Provost in 1910. He was buried in a family vault in Cowie Church-yard in Stonehaven.

Joseph McGregor Robertson (1858–1925)

McGregor Robertson graduated MA of the University of Glasgow in 1876 and MB ChM with honours in 1880. He studied physiology in Berlin under du Bois Reymond, Kronecker and Christiani and their support, with his own good record as a student, led to his appointment as Muirhead Demonstrator in Physiology at Glasgow under Professor M'Kendrick. As was then customary, he also entered medical practice. His success in practice was such that he abandoned any career in physiology and became prominent in Glasgow as an ophthalmic surgeon and as an early user of X-rays.

He wrote on many clinical topics in the later part of his career. In its early days he wrote *A Handbook of Physiological Physics* (1884) and an *Elementary Textbook of Human Physiology* (1887). He was elected a member of the Physiological Society in 1884.

Aberdeen

William Stirling (1851–1932)

William Stirling was born at Grangemouth, Stirlingshire and went to school at Alloa. At Edinburgh University he was a brilliant student, obtaining honours as he graduated BSc in 1870, DSc and MB CM in 1872 and MD with gold medal in 1875.

He travelled on the continent to work in Leipzig with Ludwig, whom he always called his master, and in Paris with Ranvier. Back in Edinburgh he was briefly assistant in natural history and then for three years assistant to Rutherford in physiology.

Although only twenty-six years of age, in 1877 he was appointed Professor of the Institutes of Medicine in Marischal College, Aberdeen and introduced there practical class teaching as in Edinburgh.

In 1885 he was elected to succeed Gamgee as Brackenbury Professor of Physiology at Manchester and he remained there until he retired in 1919, to be succeeded by A.V. Hill.

In obituary notices he is described as a very good lecturer with material presented by diagrams and models in the way he would have learnt from Rutherford, although there is no mention of elaborate lecture-demonstrations like those favoured by Rutherford (p. 200). Both in Aberdeen and in Manchester he organised practical classes which included experimental work, histology and chemistry. His writings were contributions to teaching. He translated Landois's *Lehrbuch der Physiologie des Menschen*, adding his own contributions to make *A Textbook of Human Physiology, including histology and microscopic anatomy, with special reference to the requirements of Practical Medicine*, first published in 1884 while he was at Aberdeen but continued into further editions. He also published a series of laboratory handbooks: *Outlines of Physiological Chemistry* (1881), *Outlines of Practical Physiology* (1888) and *Outlines of Histology* (1890).

He became solely a teacher and undertook no research work at Aberdeen or Manchester – indeed it was said that he discouraged research in his laboratory and anyone with leanings that way went elsewhere. Possibly he felt he had no time for research, since he was heavily involved in teaching and administrative work in the development of the University of Manchester.

He was elected a member of the Physiological Society in 1877 at the first annual general meeting, a year after the society had been formed. He held no office in the Society.

He had interests in historical matters and for the 1902 meeting of the BMA in Manchester prepared a historical memento, *Some Apostles of Physiology* dealing with medieval and later physiologists. Only a few copies were printed and distributed to friends, with the name of the recipient on the title page. The book was described and reviewed by the *British Medical Journal* at the time (*BMJ*, 1902, ii, 260–2).

Stirling died at his home in Manchester in 1932. His successor in Abderdeen in 1885 was J.A. MacWilliam, who had graduated in Aberdeen in 1880 and then worked with Schäfer at University College, London.

References

Sharpey-Schafer, E. (1927). *History of the Physiological Society during its First Fifty Years, 1876–1926*. Issued as a supplement to *Journal of Physiology*, December.

Biographies

M'Kendrick: *Proc. Roy. Soc. London B* (1926), **100**, xiv–xviii (P) (by Paton).
           Lancet (1926), i, 157–8.
Robertson: *Lancet* (1925), i, 738.
Rutherford: *Lancet* (1899), i, 538–41 (P).
           British Medical Journal (1899), i, 564–7 (P).
Stirling: *British Medical Journal* (1932), ii, 695–6.
           Lancet (1932), ii, 815–16 (P).

14

Physiologists at London medical schools

The developing institutions

Chapters 10, 11 and 12 have described the schools of physiology which after 1870 evolved in University and King's College in London and in Cambridge. The developments in these three places must be largely attributed to the high qualities of Burdon Sanderson and Schäfer; Rutherford and Yeo; Foster, Gaskell and Langley. But these men could not have achieved so much, if the institutions had not been ready for them, and indeed there were similar developments at about the same time in other disciplines. University and King's were both Colleges of the University of London, teaching medicine alongside other subjects. They had governing bodies seeking to fulfil the purpose of a university and appointments were made to their academic staffs by selection from available candidates; their physiologists were selected as physiologists. Burdon Sanderson, Schäfer, Rutherford, Foster, Gaskell and Langley were professional physiologists with no clinical appointments or responsibilities; Yeo was at first assistant surgeon at King's but in 1880 resigned this post to devote his full time to physiology.

The medical schools of the London Hospitals, and also in the provincial cities, had as yet no such established constitutions. Physiology was still taught by junior clinicians and almost without exception, as the current lecturer in physiology advanced to a higher clinical appointment, he abandoned physiology to become a clinical lecturer. Then the lectures in physiology were taken over by the next assistant surgeon or physician. They were not professional physiologists, nor salaried lecturers. In 1870–80, the lecturers in physiology were still of this type, much as they had been for the preceding forty years, but now they were forced to try and teach a more advanced physiology. The influence of Huxley, Burdon Sanderson and Foster on the University of London and the Colleges of Surgeons and Physicians had caused a new type of question to be asked by these examining bodies and they also were now requiring that the students had attended practical classes in histology, physiological chemistry and perhaps experimental physiology.

This was an awkward requirement for the medical schools. It meant that teaching laboratories and equipment were needed. It also meant that teachers of physiology needed to know and be able to teach the new experimental methods and the physiology which these methods had discovered. There is the instance of Beale (p. 42) who gave up teaching physiology at King's rather than learn the new methods.

In the medical schools of the London Hospitals, recognition of physiology as a specialist subject was delayed beyond 1885 by the conservatism of the senior clinicians who formed their governing bodies. To them lectures in physiology and other non-clinical subjects had always been the privilege and duty of young assistant surgeons and physicians and this position was maintained until in 1884 for the first time, a specialist and paid physiologist was appointed at St Mary's. However, it is apparent in the following biographies that some of these clinicians who gave lectures in physiology did have some particular interest in the subject.

A qualification for inclusion in this chapter is that the man became a member of the Physiological Society, which excludes some of the teachers in the London Hospitals. Between 1870 and 1885, physiology at the Middlesex, St Thomas's and Charing Cross was taught by physicians or surgeons, who came to some eminence in their clinical fields but made no significant contributions to the advance of the new physiology and were apparently regarded as not having sufficient interest to be asked to join the Society. Some of those listed in this chapter, in fact, made little contribution to the subject and it is not clear why they were chosen as members of the Society and others were not.

Guy's Hospital

Frederick William Pavy, FRS (1829–1911)

Pavy was born at Wroughton, Wiltshire, went to Merchant Taylor's School and in 1847 entered Guy's Hospital as a medical student. He took MB London in 1852 with many honours and proceeded MD in 1853. He served as house physician and surgeon before going to Paris to study under Claude Bernard. Burdon Sanderson and Harley were also in Paris at that time.

In 1854 Pavy was appointed Lecturer in Anatomy at Guy's and in 1856 Lecturer in Comparative Anatomy, Physiology and Microscopic Anatomy. He continued to give lectures in physiology until in 1877 he became Lecturer in Medicine. The lectures in physiology were then taken by Pye-Smith, who since 1864 had been conducting practical classes mainly in histology, but with some experimental work.

While lecturing in physiology, Pavy was becoming established as a physician. Appointed assistant physician in 1859, he gave some lectures in clinical medicine and in 1871 became full physician and lecturer in medicine, which positions he held until he retired in 1890. He had a large private practice and became recognised as a leading authority on diabetes in those days before insulin. He was much honoured by the Royal College of Physicians and served it in many capacities until his death at the age of eighty-two. He gave many important lectures and wrote many papers and books on physiological and clinical aspects of diabetes and digestion. Most important to physiologists were *A Treatise on Food and Dietetics* (1874), which remained a standard book until about 1900; *The Physiology of Carbohydrates* (1894); and *Carbohydrate Metabolism and Diabetes* (1906).

At Guy's he was noted for his fine brougham, brown with red wheels, and a pair of well-bred black horses, and also for the good food and wine of his dinner parties. His house was one of the first in London to be lit by electricity. At Guy's it was said that although he never neglected his clinical work and teaching, he excelled in neither, but was remarked for his persistence and long hours of work in his experimental laboratory. He donated a gymnasium to the students of Guy's.

About 1870, Pavy had introduced live demonstrations into his lectures in physiology, and as well as his work as a clinician, worked in experimental physiology in continuation of the period in 1854 when he was a chosen assistant with Claude Bernard. This directed his attention to carbohydrate metabolism and the physiology of digestion. At Guy's he had a small laboratory contrived under the stairs and after he retired, he worked in the laboratories of the Royal College of Surgeons and Physicians and later in the Physiological Laboratory of the University of London, South Kensington. He was elected FRS in 1863.

Pavy's work on carbohydrate metabolism covered a wide period. In his early days, he needed evidence to disprove the belief that the main site of oxidation of glucose was in the lungs and his last paper, published in the year of his death, demonstrated the uptake of glucose by all tissues unless it was injected rapidly, when some appeared in the urine. He was an active research worker after his retirement from clinical posts, publishing ten papers in the *Journal of Physiology* between 1896 and 1911. Pavy was a founder member of the Physiological Society and always a regular attender of its meetings. In 1909, on his eightieth birthday, the Society presented him with a silver bowl.

Philip Henry Pye-Smith, FRS (1840–1914)

His grandfather was John Pye-Smith, a well known nonconformist divine and his father, Ebenezer Pye-Smith, was a surgeon in Hackney, who had Pavy as a pupil. Philip Henry went to Mill Hill School and thence to University College to graduate BA in 1858. Following his father's connection, he entered Guy's Hospital Medical School with which he was thereafter always associated. He graduated MB in the University of London in 1863 and MD in 1864 with first places and gold medals. He went abroad to Vienna and Berlin and formed a particular friendship with Virchow.

His first post at Guy's in 1865 was Lecturer in Comparative Anatomy and Zoology. During the next ten years he reorganised and developed the museum in these subjects and formed friendly links with Huxley and Foster. Huxley once described him as the best educated young man in London. In 1866 he also became Demonstrator in Anatomy in succession to Pavy, with whom his career was closely linked. Pye-Smith took up the teaching of histology, then very primitive at Guy's, and in 1873 was appointed Lecturer in Physiology jointly with Pavy, becoming the sole lecturer from 1877 to 1883. He was succeeded by Golding-Bird and Wooldridge.

At the same time he was a physician–registrar in 1870 and Assistant Physician at Guy's in 1871. In 1877 he was given special care of skin diseases, in which he became an important teacher and wrote a students' textbook in this specialty. Pye-Smith had a large private practice, particularly in skin diseases. In 1883 he was elected full Physician and was appointed Lecturer in Medicine, holding these posts until retirement by age limit in 1899, when he became Consultant Physician.

A good speaker, in later life he became an acknowledged orator of the medical profession. He was in succession fellow, examiner, committeeman and censor of the Royal College of Physicians and gave several important formal lectures. In 1902 he became the representative of the College on the Senate of the University of London and was Vice-chancellor from 1903–05. He also represented the University on the General Medical Council.

At the time of the formation of the Physiological Society in 1876,

Pye-Smith was Lecturer in Physiology with Pavy and was a friend of Huxley and Foster; naturally he was a founder member of the Society. He published only one paper in the *Journal of Physiology*, in 1887 on 'Observations upon the persistent effects of division of the cervical sympathetic nerves', reporting experiments carried out at intervals in the physiological laboratory of Guy's Hospital since 1878. This was apparently his only publication in physiology, his main published works being textbooks of skin diseases and general medicine. He was a fluent linguist and in the years before the publication of Foster's textbook, he translated and summarised for his students the work of French and German physiologists. He was elected FRS in 1886.

Cuthbert Hilton Golding-Bird (1848–1939)

His father was assistant physician at Guy's and was regarded as a medical prodigy but died at the age of forty. The son went to Tonbridge School and then to King's College, London to graduate BA, before entering Guy's as a medical student in 1868. He graduated MB London in 1873 and FRCS in 1874. After a period abroad, Golding-Bird became Demonstrator in Anatomy and taught in the histology classes; he was noted for his skill in manual cutting of histological sections. In 1883 he succeeded Pye-Smith as Lecturer in Physiology, with Wooldridge, and in 1892 was appointed examiner in Physiology at the College of Surgeons. He was elected a member of the Physiological Society in 1880.

As a surgeon, he was appointed assistant surgeon at Guy's in 1875 and full surgeon and lecturer in clinical surgery in 1894. He wrote on surgical subjects but his surgical career was rather shortened when the Listerian methods replaced those of his earlier training.

He became something of a legendary figure at Guy's, particularly as a member of the United Hospitals Club and a regular and welcome attender of its dinners even in his great age. He was also noted for his hobby of collecting large numbers of intricate old clocks; it was said to be an experience to hear them all striking noon. He was also a skilled photographer and a historian of the village of Meopham, Kent, to which he retired.

Leonard Charles Wooldridge (1857–89)

His father had studied medicine at Guy's and was in practice in Overton, Hampshire. Leonard Charles was sent to the County School,

Cranleigh, and in 1874 matriculated in the University of London and entered Guy's Medical School. He took BSc London in 1877, became MRCS in 1879 and on the advice of Pye-Smith went to study with Ludwig in Leipzig. He returned to London to pass MB in 1882 and MD in 1886; he was helped towards physiology by holding the George Henry Lewes Fellowship in 1881–82 and he also worked in Berlin with Virchow. He was elected a member of the Physiological Society in 1882.

In 1884 he followed Golding-Bird as Demonstrator in Histology and Practical Physiology at Guy's and also, by the recommendation of Sir John Simon, was elected a Research Scholar of the Grocers Company, which gave him £250 per year in addition to his salary of £100 as Demonstrator. The Scholarship was continued for a second three years but he died before it ended. In 1887 he became joint Lecturer in Physiology with Golding-Bird and was also appointed assistant physician to Guy's Hospital. He died suddenly of ulcerative colitis in 1889, aged only thirty-two.

Wooldridge was a very active research worker. In his short life he published a total of twenty-six papers in Ludwig's Festschrift, du Bois Reymond's Archives, Proceedings of the Royal Society and three papers in *Journal of Physiology*, the last appearing after his death. The obituary notice in the *Lancet* includes a full list of his published papers. Much of his earlier work was done in Europe to avoid the delays and troubles which resulted from the 1877 Act. When he was settled at Guy's some work involving the use of dogs was done at the Brown Institute (p. 136).

The main subject of Wooldridge's experimental work was the clotting of blood and was summarised in his Croonian Lecture to the Royal Society in 1886 under the title of 'On the coagulation of the blood'. This dealt with the various types of fibrinogen that he had recognised in plasma, and other papers at this time dealt with circumstances under which shed blood would or would not clot. He studied intra-vascular clotting which interested him as a clinician. In the immediately following years, Wooldridge's papers on clotting were given great authority.

After his appointment as assistant physician, some of Wooldridge's published papers tended towards experimental medicine rather than physiology. One can only speculate whether Wooldridge might have become the first full-time physiologist at Guy's; more likely he would, like so many others, have become a physician. The speculation has importance in relation to Starling. Starling was a student at Guy's in Wooldridge's time, became a friend much influenced by him and to some extent succeeded him in the teaching of physiology. By Wooldridge's death it fell to Starling to develop physiology at Guy's and begin the building of an Institute there; before it was completed Starling had become Professor at University College. A further important association of the two men was that Starling married Wooldridge's widow; she became Starling's great support and encouragement.

St Mary's Hospital

Walter Pye (1853–92)

Walter was the son of Kellow Pye, honoured in the world of music, and at the age of nine went to Magdalen School, Oxford, as a chorister in the college choir. He left school with no thoughts of being a doctor but continued his general education by journeys to Spain on an expedition to observe a solar eclipse and to China, where he met Dr Manson who induced him to enter medicine.

Pye became a student at St Bartholomew's Hospital and qualified MRCS in 1876. As a student he published a paper in *Journal of Anatomy and Physiology* (1875) on the development and structure of the kidney. He filled with distinction house appointments at St Bartholomew's and became FRCS in 1877. In 1878 he was appointed to surgical posts at St Mary's Hospital, Surgeon to Outpatients and Tutor in Surgery. Also he was Lecturer in Physiology from 1878 until he resigned this position in 1883. Pye was recognised as an important young teacher in surgery, until in 1890 he was overtaken by ill health and died in 1892 of a progressive brain disease.

Pye's brief but busy career as a surgeon to outpatients resulted in a very famous book, *Pye's Surgical Handicraft* first published in 1884. Pye was working on the third edition when he was overtaken by ill health and the third edition was completed by another hand. It has continued and is now in its twenty-first, centenary, edition, still a vade-mecum of the details of surgical management of patients. Pye did a little research at the Royal College of Surgeons and in 1890 gave the Hunterian Oration on 'Rates of growth of the body with special reference to the correction of children's deformities' and in the same year published a book of similar title.

Pye joined the Physiological Society in 1879 and was often present at meetings, although he made no scientific communications.

Pye was the last surgeon teacher of physiology at St Mary's. According to Cope (1954), already it was becoming apparent that a laboratory was needed and that the lectures should be given by an expert, full-time and salaried lecturer. Matters were brought to a head in 1884 by a formal complaint from the students, who asserted that Mr Pye's teaching was in many respects at variance wtih that of the textbooks. Pye having resigned, the way was clear for the appointment of A. D. Waller.

Augustus Désiré Waller, FRS (1856–1922)

A. D. Waller must be distinguished from his father, Augustus Volney Waller (1816–70), who is known to physiologists by 'Wallerian' degeneration of nerves (p. 64) and the son always insisted on other contributions his father made. A. V. Waller lived mostly abroad and so it came about that his only son, Augustus Désiré, was born in Paris and lived his early life in Geneva. When the father died in 1870, mother and son moved to Aberdeen and A. D. obtained his medical education there, qualifying MBChM in 1878 and MD in 1881.

Following qualifications, A. D. Waller received a grant from the British Medical Association to study further under Burdon Sanderson; he had already decided not to practise but to devote himself to physiology.

In 1883, for a short time, he was Lecturer in Physiology at the Royal Free Hospital, which had the important consequence that he there met Alice Palmer, his wife, who assisted him greatly in his work and was of a rich family, of the firm of Huntley and Palmer. The announcement of Waller's engagement produced a notice on the blackboard: 'Waller takes the biscuit', to which was added 'and the tin'.

In 1884 he was chosen as full-time Lecturer in Physiology at St Mary's Hospital where he remained for twenty years. This appointment was a sign of things to come. It was made because the students at St Mary's complained that they could not pass the examinations of the University of London or the Colleges, from the lectures given by non-specialist lecturers in physiology. So Waller was given a paid lectureship with the special object of setting up practical class teaching and incidentally establishing an experimental laboratory at St Mary's. His salary was £300 per year and this was the first salaried appointment as a teacher in any of the London Hospital schools in any subject. Until then lecturers received their share of the fees paid by those who attended their courses. In connection with teaching he wrote a good students' book, *Introduction to Human Physiology* (1891). Waller was succeeded by further lecturers who made little contribution to physiology; it was not until 1920 that the first Professor of Physiology (B. J. Collingwood) was appointed at St Mary's.

In 1903 Waller noticed that in the buildings at South Kensington then being taken over by the University of London, there were vacant rooms available and he offered to turn these into a Physiological Laboratory with money mainly given by his wife's family. Thus was created a Physiological Laboratory of the University of London, with Waller as unpaid director with the status of Professor in the University of London and this continued until his death in 1922; about 1915 much argument was needed to persuade the University to continue its support.

A. D. Waller and his wife also had a large laboratory in their home, 16 Grove End Road, St John's Wood. The house had a large room built as a studio for an artist who owned the house before Waller and in this room the Wallers entertained their friends, scientific or lay. In the middle of the room was a large table on which all of the apparatus was arranged; by an ingenious arrangement of ropes and pulleys, this table could be lifted to expose a full-sized billiard table.

Waller was apparently an attractive and amusing character. He owned a series of bulldogs, really quite amiable creatures, but they did not look it. Waller was an early motorist, enthusiastic and knowledgeable about his somewhat temperamental vehicles, which he used for his ordinary affairs, including the collection of samples when he was investigating the production of CO_2 by long-distance runners. He could safely leave his vehicle with a bulldog ensconced on the seat.

Experimentally, A. D. Waller was very active and he was also a very active member of the Physiological Society. He was elected a member in 1880, four years after its foundation and between 1886 and 1920 he gave ninety communications to the Society. In these years he presided at ten meetings, either at St Mary's or the Laboratory of the University of London or in his home at St John's Wood, with the dinner in the nearby Eyre Arms Hotel. He must be the only member of the Society who has held a scientific meeting in his own house. He was Treasurer from 1896 until his death in 1922. Two sons were elected members of the Society, W.W. in 1921 and J.C., a botanist, in 1925.

Waller's research was on many topics, such as the controlled use of chloroform, the production of carbon dioxide and so the energy cost of different types of work, conductivity of the skin in emotion. He was elected FRS in 1882, while at University College with Burdon Sanderson. But most importantly he can be regarded as the man who initiated the human electrocardiogram. He had been working with electrical activity in nerve and muscle during his time with Burdon Sanderson. His first recording of the human electrocardiogram can be given in Waller's own words in a talk he later gave to students of St Mary's (Cope, 1954, pp. 81−2):

One fine day, after leading off from the exposed heart of a decapitated cat to study the cardiograms by the aid of a Lippmann capillary electrometer, it occurred to me that it ought to be possible to use the limbs as electrodes. So I dipped my right hand and left leg in a couple of basins of salt solution, which were connected to the two poles of the electrometer, and at once had the pleasure of seeing the mercury pulsate with the pulsations of the heart.

The first experiment was made in St. Mary's laboratory in May 1887 and demonstrated there to many physiologists, among others to my friend, Prof. Einthoven of Leiden, whose subsequent invention of the string galvanometer has done so much to extend and popularise the study of this subject.

Demonstrations to the Physiological and the Royal Society were on human subjects and also on Jimmy, the first of the line of bulldogs. Jimmy stood firmly in basins of saline and Gladstone had to explain to antivivisectionists who raised the question in parliament that this caused Jimmy no feeling other than might come from paddling in the sea; the look of Jimmy discouraged further investigation. Waller's contribution is now recognised at St Mary's by the Waller Cardio-Pulmonary Unit.

Willem Einthoven, FRS (1860−1927) was awarded the Nobel Prize in 1924 for his development from Waller's original observation. He had studied physics and medicine in Utrecht before becoming Professor of Physiology in Leiden in 1885. His first achievement was the string galvanometer with sufficiently rapid time characteristics. He used his physiological skill then to study the electrocardiogram of normal individuals and also cardiac irregularities, in papers published in 1907−08. He also applied similar expertise to record and study heart sounds. He was elected a foreign member of the Royal Society in 1926 and an Honorary Member of the Physiological Society in 1924. In 1925, he invited the Physiological Society to meet in Leiden, and the dinner was attended by eighty-five British and Dutch physiologists.

St Bartholomew's Hospital

In 1870 at St Bartholomew's, physiology was taught by the lectures of Morrant Baker (p. 53). He was a surgeon, a pupil of James Paget and continued to produce Kirke's *Handbook of Physiology* (p. 30) and give lectures until 1885, but he had no special training in physiology; an assistant was D'Arcy Power (see below). After 1874 Klein (p. 155) gave some of the lectures, bringing to Barts the special techniques of histology. It was not until 1893 that physiology was taught by a professional physiologist, J. S. Edkins.

In the meantime the methods of experimental physiology were used at St Bartholomew's by Lauder Brunton, Lecturer in Materia Medica. The lectures in Materia Medica were usually regarded as a chore to be undertaken by a newly appointed assistant physician in the hospital. Lauder Brunton carried out experimental investigations into the action of drugs and can be regarded as an originator in a new scientific discipline, pharmacology.

Sir Thomas Lauder Brunton, Bt, FRS (1844–1916)

Lauder Brunton was born at Hiltons Hill, Roxburgh and was educated privately before entering Edinburgh University. He graduated MBCM in 1866, BSc in 1867 and proceeded to MD in 1868 and DSc in 1870. Amongst his fellow students was Ferrier (p. 190).

In 1866–67, Brunton was house physician at the Royal Infirmary at Edinburgh under Maclagan, Professor of Medical Jurisprudence and Hughes Bennett, Professor of the Institutes of Medicine (Ch. 7). He also worked in Maclagan's recently established laboratory; Gamgee (p. 232) was at that time Maclagan's assistant. Here Brunton did the work for his MD thesis on 'Digitalis with some observations on the urine'. For some months he maintained himself on controlled diet and living conditions, measuring the volume and composition of his urine and observing the effects of increasing doses of digitalis up to toxic levels.

Helped by a scholarship, in 1867 Brunton went abroad, firstly to work in Vienna with Brücke and Rosenthal and in Berlin. He toured in Egypt and Syria before working in Amsterdam with

Kühne in 1868 and then for about six months in Leipzig in Ludwig's new laboratory (p. 134), with Kronecker as a fellow student.

In 1870 he arrived in London and was appointed Lecturer in Materia Medica at Middlesex Hospital but in 1871 moved to St Bartholomew's as casualty physician and lecturer in materia medica. He stayed at Bart's for the rest of his life, for the first twenty-four years in the post of assistant physician before in 1895 he became full physician. He retired in 1895 before the compulsory age, in order to allow promotion of younger men. He was thereafter consultant physician at Bart's.

He was regarded as a great physician, perhaps the leading physician in private practice in London at that time. He was knighted in 1900 and made a baronet in 1908. He served in many offices at the Royal College of Physicians and delivered many formal lectures. In the later stages of his career he became interested in the question of physical fitness, following reports of the low standard of army recruits. He took part in the formation in 1905, and was for a time secretary, of the National League for Physical Education and Improvement.

After about 1890 he ceased active experimental research work but his work as a younger man had virtually begun the science of pharmacology, which for the next fifty years became an important section within physiology. In Edinburgh and in his visits to European laboratories, he had carried forward his accurate description of the actions of digitalis. As a house physician in Edinburgh he had studied the clinical value of amyl nitrite in angina pectoris and he studied its actions. Much of this early work depended on the measurement of arterial blood pressure by mercury manometer in animals and by sphygmograph in man. By his experience in Edinburgh and abroad, when he went to London Brunton was probably the best trained experimental physiologist in the city and was immediately asked to join Burdon Sanderson, Foster and Klein in writing *Handbook for the Physiological Laboratory* (see p. 133). At St Bartholomew's Hospital he found a small room, 12 feet x 6 feet, used for washing jars, which he turned into a laboratory, paying the expenses himself and providing a salary for an assistant. New building in the college in 1881 provided a better laboratory.

From 1870 to 1890 he published many papers describing his experimental results with therapeutic agents, using the methods of experimental physiology. In relation to his clinical work he tried to establish a logical basis for any medical treatment, comparing the pathology of a disease with the action of drugs which might correct it. The book in which this approach was presented was *Textbook of Pharmacology, Therapeutics and Materia Medica* first published in 1885 and dedicated to 'Carl Ludwig, my beloved master'. Other books were: *Pharmacology and Therapeutics* (1877, dedicated to Burdon Sanderson); *On Disorders of Digestion* (1896, dedicated to Gamgee); and *Therapeutics of the Circulation* (1908, dedicated to Kronecker). These are more precise and logical than previous books such as that of Ringer (1869, p. 153). Brunton's textbook was the forerunner of later books on pharmacology such as Cushny's *Pharmacology and Therapeutics* (1899) and Clark's *Applied Pharmacology* (1923).

Brunton's scientific work was quickly recognised by his election as FRS in 1874, when he was only thirty years old. He subsequently served on the Council of the Royal Society and was its Vice-President in 1905–06. He was in 1876 a founder member of the Physiological Society, being a member of the group which drew up the rules. He presided at the second meeting and on several occasions demonstrated apparatus at scientific meetings.

Sir D'Arcy Power (1855–1941)

D'Arcy, son of Henry Power (p. 57), was educated in London before entering New College, Oxford in 1874 to read natural history, in which he obtained Honours in 1878. He proceeded to St Bartholomew's Hospital for his medical education, qualifying MB in 1882. In 1883 he became a Fellow of the Royal College of Surgeons. He had kept some connection with Oxford acting as Demonstrator in Physiology about 1878 and taking his MA in 1881.

His main career was as a surgeon at Bart's, being House Surgeon to Savory (p. 53) in 1882, Curator of the Museum in 1884, Assistant Surgeon in 1898 and Full Surgeon in 1904. He retired in 1920, when he became consulting surgeon and governor. He was an eminent surgeon, twice vice-president of the Royal College of Surgeons and also widely known amongst doctors of the time, as their examiner in physiology or surgery at many Universities or Colleges.

D'Arcy Power was also widely known as a writer on historical matters connected with surgery and particulary as a biographer. He completed and brought to publication Plarr's *Biographical Notices of the Fellows of the College of Surgeons* and he wrote many notices for the *Dictionary of National Biography*. He collected together all the known portraits of William Harvey; he founded a Samuel Pepys Club and wrote a medical history of Mr and Mrs Samuel Pepys.

He was knighted in 1919 for his services to the military hospitals in London during the 1914–18 war.

Amongst his activity as a surgeon and historian, he did a little work in physiology. His first publication was with Lauder Brunton on albumin in urine. In his early years as a surgeon at St Bartholomew's Hospital he was assigned to the duty of Assistant Demonstrator in Physiology, assisting Baker (p. 53). In 1880 he combined with Harris to write a *Manual for the Physiological Laboratory*, and in the same year was elected a member of the Physiological Society. In 1890–1903 he joined his father (p. 58) at the Royal Veterinary College as Lecturer in Histology and Physiology.

Westminster Hospital

Sir William Henry Allchin (1846–1912)

Of an old Kentish family and the son of a doctor in Bayswater, William Henry Allchin happened to be born in Paris. He was educated privately and in 1865 entered University College, London to study medicine. He qualified MRCS and LSA in 1869 and immediately served as medical officer on *SS Great Eastern*, when that mammoth ship was engaged in cable-laying.

In 1871 he was back in London and passed MB (London) with honours. He was assistant physician at the Western Dispensary, St Marylebone Dispensary and the Victorian Hospital for Sick Children; the last of these connections he retained for the rest of his working life. At this time he was also Lecturer in Comparative Anatomy and Animal Physiology at University College. He retained this post for some time after he went to the Westminster and was later elected a life Governor of University College.

Allchin's life was spent as a physician at the Westminster Hospital. First in 1872 he was appointed Medical Registrar and Demonstrator in Practical Physiology. In 1873 he became Assistant Physician and in 1877 Physician to the hospital. He was a good and much loved clinician and teacher who served the hospital well until he retired in 1907; he was Dean 1878–83 and 1890–93.

He lectured in the medical school in several capacities: Demonstrator in Histology and Practical Physiology from 1872–78, Lecturer in Pathology from 1873 to 1889, Lecturer in Anatomy and

Physiology, 1878—82 and Lecturer in Medicine from 1882—92. At the time he was Lecturer in Physiology, the course in histology and practical physiology was given by Murrell and then North (see below).

Allchin did no experimental work in physiology. His published work was entirely on clinical topics. He contributed a section on disorders of digestion to *Quain's Dictionary of Medicine* (p. 88), and on the peritoneum to *Allbutt's System of Medicine*. In 1900—03 he published his own *Manual of Medicine*.

Allchin was active in the Royal College of Physicians, giving several of the formal lectures and becoming Assistant Registrar about 1883. At that time there was much discussion and many proposals concerning the relationship between the University of London, the London Medical Schools and the Royal Colleges in the granting of degrees and registrable medical qualifications. Because he differed from the opinion of the majority in the College of Physicians, Allchin resigned as Assistant Registrar but continued to take a prominent part in the debate. Eventually when in 1900 the University of London had been reconstituted, Allchin and Pye-Smith (p. 211) were appointed the representatives of the Royal College of Physicians on the new Senate and the College gave Allchin a vote of thanks and the sum of 100 guineas in recognition of his work.

Allchin was knighted in 1907 and in 1910 was appointed Physician Extraordinary to King George V.

He was elected a member of the Physiological Society in 1877 at the first Annual General Meeting. He frequently attended meetings but held no office and made no communications to the Society.

William Murrell (1853—1912)

Murrell was a medical student at University College, London, qualifying MRCS and LSA in 1874. He was then appointed Demonstrator in Physiology at University College and Sharpey Scholar, thereby assisting Burdon Sanderson and Schäfer in their classes. He also assisted Ringer (p. 155) with some of his experiments.

In 1877 he followed Allchin to the Westminster Hospital as Medical Registrar with also the assignment to teach histology and practical physiology. In 1886 he became Assistant Physician and in 1896 Physician to Westminster Hospital.

In the hospital he lectured on pharmacology and on materia medica and therapeutics, these being his own interests and in which he had benefited from contact with Ringer. Murrell wrote two very successful books – *What to do in cases of poisoning* (1891) and *Manual of Pharmacology and Therapeutics* (1896).

He was never a member of the Physiological Society, despite his connections with University College. He seems to have been of somewhat retiring nature.

William North (1854–?)

North's father was a surgeon in York and William was educated at St Peter's College. He matriculated in 1874 as a pensioner at Sidney Sussex College, Cambridge and passed second class in the Natural Sciences Tripos of 1877. He graduated BA in 1878 and MA in 1888. In 1878, from Cambridge, he published a paper in the first volume of the *Journal of Physiology* on the effect of starvation on the excretion of urea.

From 1879–81 North was Sharpey student at University College, London and in 1881–82 became joint Lecturer in Physiology at the Westminster Hospital Medical School. In 1883 he was appointed Research Scholar in Sanitary Science of the Grocers Company. In 1884, North applied unsuccessfully for the new Chair in Physiology in Leeds (p. 238).

The Grocers Company renewed his scholarship for two further years, allowing him to return to Italy to continue his study of malaria. The results appeared as a book: *Roman Fever: the results of an enquiry during three years residence on the spot into the origin, history, distribution and nature of the Malarial Fevers of the Roman Compagna* (1896).

He was elected to the Physiological Society in 1880 and gave three demonstrations to the Society. He then developed other interests as antiquarian and author. It is recorded that in 1897–98 North was technical manager of a firm in Paris engaged in colour photography. He took part in a Board of Trade enquiry into the cost of living and in 1914–19 was translator to the Ministry of Food and later the Bureau of Entymology.

Charles Douglas Fergusson Phillips (1830–1904)

His father was a distinguished officer in the Peninsular and Waterloo Campaigns. Charles Phillips graduated LLD (Aberdeen) and MD (Marischal College) in 1859. He became FRCS (Edinburgh) in 1866 and was also a Fellow of the Royal Society of Edinburgh.

After qualifying, he began medical practice in Manchester until in 1867 he moved to London. There he obtained a large practice and in 1875 was appointed Lecturer in Materia Medica and Therapeutics at Westminster Hospital. He was also examiner in this subject at the Universities of Edinburgh, Glasgow and Aberdeen.

In 1878 he was forced to give up practice by paralysis resulting from injuries in the Nine Elms Junction railway accident, for which he eventually was awarded £15,000 damages on the grounds that this was his annual income at the time of the accident. Nevertheless in 1883 he could resume some work. In his later life he was a man much respected and prominent in Conservative circles; he was asked, but declined, to stand as Parliamentary candidate for the Universities of Glasgow and Aberdeen.

Phillips wrote several books on materia medica, the most important being *Materia Medica and Therapeutics* first published in parts in 1874–92 and reaching a third edition in 1903. In 1901 with M. S. Pembrey, Phillips published *The Physiological Action of Drugs*, which was a manual of experiments for elementary students. Phillips became a member of the Physiological Society in 1885.

The London Hospital

Jeremiah McCarthy (1837–1924)
McCarthy was born in Dublin and entered Trinity College, Dublin in 1854 aged seventeen. He became MA in 1863 and then went to London to study medicine at the London Hospital Medical School, qualifying MRCS in 1866 and LSA in 1867. He graduated MB London in 1868 and became FRCS by examination in 1873.

He entered the surgical staff of the London Hospital, being Assistant Surgeon 1870–77, Surgeon 1877–98, and then Consultant Surgeon. From 1877 to 1889 he was Lecturer in Physiology in the London Hospital Medical School and Lecturer in Surgery 1890–93. He was a founder member of the Physiological Society in 1876, and a frequent attender at meetings.

McCarthy's publications were on surgical subjects, and apparently he made no contributions to physiology. McCarthy's successor as Lecturer in Physiology was another surgeon, C. W. Mansell-Moullin (1851–1940); when he resigned in 1895, the London appointed its first specialist physiologist, Leonard Hill.

St George's Hospital

Herbert Watney (1843–1932)

Herbert Watney was a member of the brewing family, although he himself became a strict teetotaller. He was educated at Rugby and in 1862 proceeded to St John's College, Cambridge, where he was a noted oar, rowing in the Cambridge Boat in three successive years and winning the Colquhoun Sculls. He took his BA in 1866, MA 1869 and then qualified MRCS in 1876.

Watney's medical education was at St George's Hospital where, after the usual house appointments, he was elected Assistant Physician in 1877. As a junior clinician, Watney was also Demonstrator in Physiology (1873–77), Lecturer in Materia Medica (1878–79), Lecturer in Physiology (1881–83). He became full physician in 1881 and was elected FRCP in 1883, when he was also Examiner in Medicine in Cambridge. He joined the Physiological Society in 1881; for a period he was engaged in research on the thymus with Burdon Sanderson at University College and Klein at the Brown Institute.

However, he purchased a large estate in Berkshire and in 1883 abandoned his clinical practice and his work in physiology. In agricultural circles he became regarded as an expert on afforestation and pedigree cattle and his own estate became a model, particularly by his installation of piped water supplies to all its villages.

Sir Edward Charles Stirling, FRS (1848–1919)

E. C. Stirling was born in South Australia at Strathalbyn and was educated at St Peter's College, Adelaide. In 1864 he left Adelaide and spent a year in France before entering Trinity College, Cambridge in 1865. He passed with second class honours in the Natural Sciences Tripos of 1869 and went for medical training to St George's Hospital, London. He qualified MRCS in 1872, took his MA and MB, Cambridge in 1873–74 and in 1874 also became FRCS (England).

Stirling entered his medical career at St George's firstly as house surgeon and in 1879 was appointed assistant surgeon. He was demonstrator in Physiology in 1878–79 and Lecturer in Anatomy and Physiology in 1880. In 1877 he had travelled back to Adelaide and married an Adelaide girl. They returned to London for Stirling to continue his career as a surgeon at St George's and he took his MD Cambridge. However they returned to Adelaide in 1881 at a time when medical education was just beginning in that city. Stirling was appointed Lecturer in Physiology in the newly founded University of Adelaide and also Honorary Medical Officer at the Adelaide Hospital.

His activities then became widespread. In addition to his University and surgical work, Stirling began an extensive study of the anthropology of the aborigine, he was a keen gardener, and was elected a member of parliament. He became Director of the South Australian Museum, which became a memorial to his work, he himself contributing much material to its sections of ethnology and palaeontology. He was something of an explorer; his ride across Australia with Kintore gaining him the honour of CMG, and he was also a member of the Horn Expedition into Central Australia in 1894, when he made important contributions as its anthropologist and was also its medical officer. By this general scientific activity, particularly some of his work in palaeontology, he gained the recognition of FRS in 1893.

Stirling was elected a member of the Physiological Society in 1892 but was never an active physiologist, having neither the aptitude nor the opportunity for experimental work. He taught general biology but saw the requirements to establish a medical degree in Adelaide. He persuaded a wealthy pioneer, Sir Thomas Elder, to endow a Chair of Anatomy and so started the medical school in 1885; Stirling also used his influence to persuade his medical friends to undertake the further courses of lectures. In 1890 his own Lectureship was changed to a Professorship. He himself did a little surgery, and produced meticulous description of some of his cases. He was the undoubted leader of the profession of Adelaide at that time and spent much time and skill in the organisation of the new medical school. His main contribution to medical literature was an article with Verco on Hydatid Disease in *Allbutt's System of Medicine*; at that time hydatids were common in South Australia.

As a student and ardent preserver of the local flora and fauna his garden at Mount Lofty became a showplace. He was a good shot, rode well on horse or camel and had the stamina required of an outback traveller.

Clinton Thomas Dent (1850–1912)

After schooling at Eton, Dent entered Trinity College, Cambridge in 1868 and took his BA in 1873 and MA in 1877. He then went to St George's Hospital to qualify MRCS in 1875 and obtained FRCS in 1877. Although he never completed his medical graduation at

Cambridge he was many years later honoured by the degree of Master of Surgery, *honoris causa*.

Immediately after qualifying Dent was a house surgeon at St George's and after 1877 undertook various teaching duties — Demonstrator in Anatomy (1877), Lecturer and Demonstrator in Physiology (1881–83). He was appointed assistant Surgeon in 1881 and full Surgeon in 1895.

Dent became an important figure in the Royal College of Surgeons; he was Hunterian Professor in 1908, examiner 1902–11 and Vice-president 1911. In 1904 Dent was chosen as Chief Surgeon to the Metropolitan Police and devoted much time to the study of the injuries suffered by police constables in the course of their work. The police force, officially and by its individual members, paid their respects at his funeral.

Dent was elected a member of the Physiological Society in 1893 but his writings were entirely on surgical topics; his only contact with physiology was at the time of his lectureship.

To the general public Dent was known as a mountain climber, in succession Secretary, Vice-president and President of the Alpine Club. He achieved many first ascents in the Alps and Caucasus Mountains.

William Ewart (1848–1929)

The son of an English father and French mother, William Ewart was educated partly in England and partly in Paris, where he obtained the University diploma of *Bac. ès Lettres*. For medical education he entered St George's Medical School in 1869 but interrupted his studies to work in the medical service of the French Army during the Franco-Prussian war. He then passed MRCS in 1871 and LRCP in 1873 and for a time was House Physician at St George's. In 1873 he entered Caius College, Cambridge, obtained first class in the Natural Sciences Tripos of 1876 and was also House Physician at Addenbrooke's Hospital. He took the degree of MB Cambridge in 1877 and went to Berlin for postgraduate experience. In 1877 he was back at St George's firstly as Curator of the Museum and Demonstrator in Pathology; in 1881–83 he gave lectures in physiology and physiological chemistry.

His clinical work in the hospital began by his becoming assistant Physician in 1882 and he became full Physician from 1887 to 1907. He was also assistant Physician at the Brompton Hospital for Consumption and Diseases of the Chest. He is described as a clever but eccentric physician who acquired no large consulting practice, but his students gained amusement and instruction from his teaching and flow of ideas. His published works were on diseases of the chest. He became a member of the Physiological Society in 1886.

References

Besterman, E. and Creese, R. (1979). 'Waller – pioneer of electrocardiography', *British Heart Journal*, **42**, 61–4.
Cameron, H. C. (1954). *Mr Guy's Hospital 1726–1948*. London: Longmans Green.
Cope, Z. (1954). *The History of St. Mary's Hospital Medical School*. London: Heinemann.

Biographies

Allchin: *British Medical Journal* (1912), i, 402–5.
Brunton: *Lancet* (1916), ii, 572–5 (P).
                    British Medical Journal (1916), ii, 440–2 (P).
Dent: *Lancet* (1912), ii, 730–3 (P).
Einthoven: *Proc. Roy. Soc. London B* (1927–8), **102**, v–viii (P).
Ewart: *Lancet* (1929), ii, 408–9 (P).
Golding-Bird: *British Medical Journal* (1939), i, 590–1 (P).
McCarthy: *The Medical Directory* (1923). London: Churchill.
Murrell: *Lancet* (1912), ii, 124.
North: *Alumni Cantabrigienses*, Part II, Vol. IV (J. A. Venn, 1951). Cambridge University Press.
Pavy: *Lancet* (1911), ii, 977–80 (P).
Phillips: *Lancet* (1904), ii, 1619.
Power, D'A.:   *British Medical Journal* (1941), i, 836–7 (P).
Pye: *Lancet* (1892), ii, 916.
                    Cope (1954).
Pye-Smith: *British Medical Journal* (1914), i, 1215–16 (P).
Stirling: *British Medical Journal* (1919), i, 785.
Waller: *British Medical Journal* (1922), i, 458–9.
                    Proc. Roy. Soc. London B, (1922), **93**, xxvii–xxx (by Halliburton).
                    Cope (1954) (P).
                    Besterman and Creese (1979) (P).
Watney: *Lancet* (1932), i, 1231.
Wooldridge: *Lancet* (1889), i, 1281–2.

15

Physiologists at provincial medical schools

The developing institutions

In the Provincial as in the London Medical Schools (Ch. 14), in 1870–80 physiology was still taught by clinicians with the same difficulties of equipment and inadequate teaching; the University of London, the College of Surgeons and Society of Apothecaries were also the examining bodies for students from the provincial schools.

In the provinces a new factor was introduced when the medical school of a city combined with an existing college of further education, to form a university-type college able to seek its own charter and ultimately reach university status and grant medical degrees forming a registrable qualification. The time of union between the medical school and college was very favourable for the new appointment of a Professor of Physiology and the provision of laboratories for experimental work and research.

This first occurred in Manchester, where in 1872 the Royal Medical School merged with Owens College. Immediately Gamgee was appointed Bracken-bury Professor of Physiology, having been selected by an appointing committee of Owens College (Thompson, 1886). In Liverpool in 1878, when the Royal Infirmary School of Medicine merged with the University College of Liverpool, Caton was given the title of Professor of Physiology (Kelly, 1981). In Leeds, the Medical School joined with Yorkshire College in 1884 and a Professor of Physiology (Birch) was immediately appointed (Anning and Walls, 1982).

In Birmingham, until 1880 all medical teaching was given in Queen's College of Birmingham. Medical teaching had however become stagnant, when in 1880 Mason College of Science was founded and in 1881 an arrangement was made by which a Professor of Physiology (Haycraft) was appointed by Queen's College to teach physiology in Mason College to students of both colleges. It was much later that the bulk of the medical teaching in Queen's College became the Queen's Faculty of Medicine in Mason College and later the University of Birmingham (Vincent and Hinton, 1947).

Manchester

Sir William Roberts, FRS (1839—99)

William Roberts was born at Bodedern, Anglesea where his father combined the work of farming with that of the only medical man in the neighbourhood. His was a Methodist family, several members of which became distinguished in Manchester, one of William's brothers becoming Lord Mayor. After early education in Manchester, William attended Mill Hill School in London and in 1849 entered University College, London as a medical student. He had an outstanding career, gaining many honours and prizes as he graduated BA in 1851, qualified MRCS and LSA in 1853 and graduated MB of the University of London in 1853. He took MD in 1854. Such a brilliant student inevitably attracted the interest of Sharpey, who gave Roberts a lasting interest in physiology. After graduating, Roberts travelled on the continent.

When he returned in 1854, William Roberts went to Manchester as House Surgeon in the Royal Infirmary and very soon in 1855 the young man of only twenty-five was appointed full physician and began lecturing in the Manchester Royal School of Medicine. He lectured first in anatomy and physiology (1856—57), then in pathology and morbid anatomy (1857—63) and after 1863 in medicine. The School of Medicine in 1872 amalgamated with Owens College and Roberts was thus the first Professor of Medicine in the College and later in Victoria University. For thirty years, Roberts was an energetic, thorough and lucid teacher of medicine in Manchester.

At the same time he became eminent in the Royal College of Physicians of which he became a Member in 1860 and a Fellow in 1865. He delivered the Goulstonian Lectures (1866), the Lumleian Lecture (1880), Croonian Lecture (1892) and Harveian Oration (1897). Also he served on the Council (1882—84) and was Censor (1889—90). The subjects of his lectures show his main scientific and medical interests. In 1866 and 1892 his subject was 'Use of solvents in the treatment of Urinary Calculi and Gout', based on experiments carried out in his private laboratory; this was at the time the most advanced study of the metabolism of uric acid. This work can be linked with an important book also containing much of his own research — *A practical treatise on Urinary and Renal Disease*, first published in 1865 and reaching a fourth edition in 1885. Physiological material from this book was cited in Foster's *Textbook of Physiology* in 1877. Another topic in his lectures of 1885 was 'Digestive Ferments and Artificially digested Foods' and this too contained much of his own research and was brought up to date in *Collected Contributions on Digestion and Diet* published in 1891.

Roberts was elected FRS in 1877 and served on the Council in 1890–91. Important papers appeared in the Proceedings of the Royal Society, for example on the staining reactions of red corpuscles and on biogenesis. A practical result of his work on this latter topic was much new information about the sterilisation of fluids and the effect of ambient temperature on bacterial growth at a time when bacteriology was in its infancy. In 1885 recognition of his scientific and public work came in an unsought knighthood; his biographers state that his first intimation of this preferment was a paragraph in a Manchester newspaper.

In 1889 Sir William Roberts left his high position in Manchester and went to London, chiefly to obtain relief from the calls of a large consulting practice. In London he continued administrative work. He was elected a Fellow of the University of London and he became an active member of the Committee managing the Brown Institute, succeeding Sir Richard Quain as its chairman in 1897. He represented the University of London on the General Medical Council and took great interest in the whole question of university teaching in London. He served on committees dealing with opium of which he had seen the effects in India and on the effects of moisture in the weaving sheds on the health of workers, of which he had had wide experience in Manchester. Roberts' election to the Physiological Society in 1890 was at this late stage of his career when he was living in London and many years after his first beginning physiological-style research in Manchester. His research had been in the old-fashioned way in his own home. Even when a laboratory became available to him in Owens College, he used his own laboratory.

During the last twenty years of his life, Roberts spent as much time as possible at his country residence, Bryn near Llanymawddwy, Merionethshire where he took great interest in developing his estate, providing his tenants with model farmhouses. He studied the flora of the district and enjoyed fishing in the River Dovey flowing past his house. He was buried in Llanymawddwy at a ceremony attended by official representatives and colleagues from Manchester and the families of his tenants with children carrying primroses.

Arthur Gamgee, FRS (1841–1909)

Arthur Gamgee was born in Florence where he lived for the first ten years of his life. His father was Joseph Gamgee, a veterinary surgeon of Edinburgh, distinguished for his researches particularly in regard to rinderpest. Arthur was educated at University College School, entered medical education at Edinburgh, graduating MD in 1862 with a gold medal for his thesis on foetal nutrition.

He was immediately appointed Lecturer in Physiology at Surgeon's Hall, Edinburgh and Physician to the Royal Hospital for Sick Children. From 1863–69 he was assistant to Maclagan, Professor of Medical Jurisprudence at Edinburgh University and in his laboratory began researches on haemoglobin; several papers were published and in 1872 he was elected Fellow of the Royal Society. During 1871 he had gone to Heidelberg and Leipzig to work with Kühne and Ludwig.

In 1873 Gamgee was appointed the first Brackenbury Professor of Physiology at Owens College, Manchester. In the previous year the Royal Manchester School of Medicine had amalgamated with Owens College. During the twelve years that Gamgee was in Manchester, Victoria University was begun, with Owens College at first its only constitutent college and Gamgee, as Dean of the Faculty of Medicine, took his full part in this development. He was also physician to the Hospital for Consumption and as a lecturer contributed to the activity of Owens College for providing general lectures for groups of working-class people; Huxley was another lecturer in these courses. In 1882–85 Gamgee was also Fullerian Professor of Physiology at the Royal Institution in London.

However Gamgee's main work in Manchester was to establish a department of physiology, attracting more students by the high quality of his lectures and teaching and by his own research work. His efforts were recognised by Michael Foster; in founding the *Journal of Physiology* in 1877, Foster asked Gamgee to be a co-editor as one of the few then actively engaged in physiological research. He was a founder member of the Physiological Society in 1876.

It is not clear why in 1885 Gamgee left Manchester and entered private practice as a physician at St Leonards-on-Sea; he was appointed Assistant Physician at St George's Hospital in London and Lecturer in Pharmacology and Materia Medica. Ill-health intervened, and after a period in which he did research work in Cambridge, Gamgee went to Switzerland, at first at Berne, then Lausanne until he settled at Montreux where he formed a lucrative private practice.

He fitted up a laboratory in the basement of his house where he continued his researches. He often visited his friend Kronecker in Berne and in England worked in the laboratory in Cambridge and in the laboratory of the University of London, founded by A. D. Waller (p. 216). In 1904 he was asked by the Carnegie Institute to advise on the methods of conducting research in metabolism and calorimetry, for which he twice visited America and toured European centres. He gave up his practice in Montreux and himself worked on this topic in Cambridge. Gamgee's last published paper (1908) was the description of a method of continuously recording body temperature but he never published results using the method. He died of pneumonia in Paris on a trip to visit his friend Kronecker in Berne. He was buried in a family vault at Arno's Vale Cemetery, Bristol.

In 1908 he was honoured by the award of the degree of DSc in the University of Manchester, the citation saying: 'In Arthur Gamgee we welcome an old colleague who for many years illustrated the Chair of Physiology in Owens College, and whose services in the establishment of the former Victoria University and in securing its medical charter are not lightly to be forgotten.' To physiologists however the importance of Gamgee lies in his establishment there of an experimental laboratory and his institution of teaching based on experimental work.

His own research was original in that he was one of the first physiological chemists. In his first phase in Edinburgh he described methaemoglobin produced by the action of nitrites and he also described accurately and rather completely, cystine. He continued to work on haemoglobin and in 1902 gave a Croonian Lecture to the Royal Society entitled 'Certain Chemical and Physical Properties of Haemoglobin'. In Schäfer's *Textbook of Physiology* (1898) he wrote a chapter on haemoglobin. During his period in Manchester his interest and understanding of physiological chemistry led him to write *Textbook of the Physiological Chemistry of the Animal Body, including an Account of the Chemical Changes occurring in Disease*, published in 1880. This was the first account of English of chemical investigation of the phenomena of life and remained an important book for the next thirty years. A second volume dealing with digestion did not appear until 1893 and incorporates work he had done in Kronecker's laboratory at Berne and in his own laboratory at Montreux. He also reported experiments in other fields of physiology such as the action of the vagus on the heart and finally the continuous recording of body temperature.

As a young man in Edinburgh and for the twelve years in Manchester, Gamgee was an effective lecturer and teacher of medical students and introduced his classes to experimental work in physiology. Finding a dearth of suitable textbooks he translated from the German Hermann's *Human Physiology*; published in 1875, this translation was before Foster's textbook and was sufficiently widely accepted to require a second edition in 1878.

Biographers, such as M'Kendrick and Schäfer in *British Medical Journal*, regarded Gamgee with great affection and as a man of singular versatility and brilliance; they obviously thought that his originality had inadequate recognition by his contemporaries. He had a fiery enthusiasm which often carried him too far in developing his scientific arguments and in his personal relationships with colleagues. The *Lancet* obituary notice concludes: 'Of a most affectionate disposition and of sanguine temperament, he may not have always chosen to display these qualities to others.'

John Priestley (?–1940)

Priestley's career was entirely in Manchester. In 1874 he was awarded the Platt Studentship in Physiology at Owens College and remained with Gamgee to become assistant Lecturer in Physiology and Histology from 1877 until 1880. He was elected a member of the Physiological Society in 1878.

Priestley obtained medical qualifications, LSA and MRCS England in 1882 and was House Surgeon at the Manchester Royal Infirmary and assistant Physician at the Manchester Hospital for Consumptives. He entered the school medical services, becoming Senior School Medical Officer for the Staffordshire County Council until his retirement in 1920. After retirement he lived in Bridgnorth until he died in 1940.

In his brief period as a physiologist, Priestley published papers on three topics. In 1877 with Gamgee and Larmuth there was a paper in *Journal of Anatomy and Physiology* on the toxicity of phosphoric acids; in *Journal of Physiology*, Vol. 1, 1878 he wrote two papers on Batrachian lymph hearts and in the same volume a paper with Gamgee on the effect of stimulation of the vagus on the heart of the anaesthetised dog. These, which seem to be Priestley's only published work in physiology, have the distinction of including the first three papers published in *Journal of Physiology*.

William Horscraft Waters (1855–87)

Waters was born and educated in London with the first intention of a career as an architect. However he showed a preference towards scientific subjects, particularly chemistry, and in 1875 entered Christ's College, Cambridge with an open scholarship to read natural sciences. He passed first class in the Natural Sciences Tripos of 1878 and was immediately appointed Demonstrator in Physiology under Michael Foster; in 1879–80 he went to Germany to study under Kühne and Ludwig, carrying out research on the vasomotor nerves of the frog, published in the *Journal of Physiology* in 1885.

In 1882 Waters was appointed Senior Demonstrator in Physiology at Owens College, Manchester under Gamgee and at the same time he joined the Physiological Society. He was a successful teacher and in 1885 was promoted to Assistant Lecturer in Histology. In 1885–86, during the interval between the resignation of Prof. Gamgee and the arrival of his successor, Stirling, Waters gave the full course of lectures on physiology and ran the department. He was beloved and esteemed by students and staff and his sudden death was a great loss. He had great aptitude for the construction of ingenious pieces of apparatus and wrote a small book, *Historical Notes for the use of Medical Students*.

Liverpool

Richard Caton (1842–1926)

Although Richard Caton was born in Bradford, he was of a Lancastrian family connected with Heysham and Caton. He went to school in Scarborough and received his medical education in Edinburgh, passing MB in 1867 and taking his MD in 1870 by a thesis on the migration of leucocytes. He was thus a student under Hughes Bennett.

After 1868 he was settled in Liverpool to practise as a physician with appointments at the Liverpool Royal Infirmary; in its School of Medicine he was at first Lecturer in Comparative Anatomy and Zoology. From 1872 until 1891 he lectured in physiology in different capacities. In 1872 he was Lecturer in Physiology at the Liverpool Royal Infirmary School of Medicine. In 1878, this school was merged with the University College of Liverpool and Caton became the first Professor of Physiology in the College; he remained Professor until 1891 when he retired to be succeeded by Gotch, who was the first full-time Holt Professor of Physiology. By then the College had become the separate University of Liverpool. Caton was an imporant figure in these developments and after his retirement as Professor, he later represented the University on the General Medical Council and was for a period Pro-Chancellor of the University of Liverpool (Kelly, 1981).

In 1873 Caton gave the opening address at the Liverpool Royal Infirmary School of Medicine, being then described as Lecturer in Physiology. Much of his address dealt with public health and hygiene in which he was much interested. This was also the occasion of the opening of a new extension of the school containing laboratories for the teaching of physiology, and Caton said that 'for its size there is no school in the country more fully equipped for the work of medical teaching in all its scientific and practical departments'. Caton was an administrator rather than a practical physiologist, although he did some work on cerebral localisation in the new laboratories; certainly he handed on to Gotch in 1891 an established laboratory. He was a founder member of the Physiological Society.

Caton achieved considerable status as a physician in practice and in the Royal Infirmary. He was particularly interested in diseases of the heart, wrote a book, *Prevention of Valvular Disease of the Heart* and was active in founding a new Heart Hospital in Liverpool.

In 1907 Caton became Lord Mayor of Liverpool. This was the culmination of his activity in civic affairs, particularly in relation to public health. He was a considerable classical scholar and wrote several works on archaeology in Greece and Egypt.

When Caton died in 1926, writers of obituary notices attached little significance to his research work; retrospectively, it can be seen to have been a significant beginning of the electroencephalogram (Cohen, 1959). About 1874 with a grant from the British Medical Association, Caton tried to record electrical activity by electrodes placed on the exposed surface of the brain of mammals. The apparatus was inadequate but Caton did find some electrical activity and some additional activity evoked by peripheral stimulation. Although he discussed the publication of his results with Burdon Sanderson, only meagre accounts appeared (Cohen, 1959). However, Berger referred to Caton's work in the introduction to his classical paper of 1929 which is regarded as the beginning of electroencephalography.

Leeds

From 1871 to 1883, lectures in physiology were given by C. J. Wright.

Charles James Wright (1842–1908) was a Yorkshireman, born at Wakefield and educated at St Peter's School, York. His medical education was in the East Parade days of the Leeds School of Medicine and at Guy's Hospital, to qualify MRCS and LSA in 1864–65. From 1865 he was in general practice in Leeds, where a special skill in obstetrics became recognised until he was accepted as the best consultant; after 1884 he was firstly Lecturer then Professor of Midwifery and Obstetrics until his retirement in 1907, having served the Medical School for forty-one years. He joined the Council of the Medical School in 1870, undertaking the duty of lecturing in physiology, general anatomy and pathology. He had no training as a physiologist and no experience of experimental work.

For a short period in 1883–4 Wright was assisted by

Ernest Henry Jacob (1849–94), who had graduated at St Thomas's Hospital before he set up as a physician in Leeds in 1875. In 1881 he joined the Council of the Medical School and acted as Demonstrator in Physiology. In 1884, with the amalgamation with Yorkshire College he was appointed part-time Professor of Pathology until his death in 1894. He was Assistant Physician, then full Physician in the Infirmary. A man of wide interests, he was an authority on the heating and ventilation of houses.

Before Jacob two other clinicians, James Walker and John Horsfall, had been Assistants in Physiology to Wright, Walker from 1871 to 1883 and Horsfall from 1874 to 1883. They gave some lectures and had charge of practical classes in histology. Neither lectured in other subjects after they retired from teaching physiology.

Although they had no training in physiology, Wright, with Jacob and other assistants, began some improvement in the teaching of physiology. The growing importance of the subject had led in 1869 to a separate course of lectures and a prize medal separate from anatomy, and some practical histology was introduced. At first this was only the setting up of microscopes in the lecture theatre, displaying already mounted specimens which were changed each day. Next was added the attendance of a demonstrator on Thursday evenings to give instruction in the use of microscopes; there was at that time no separate class room for histology. In 1876, after the Yorkshire College of Science had been opened, the classes in chemistry were held in

Yorkshire College, freeing the one laboratory in the Medical School for use as a physiological laboratory. Thus in 1876–77 the prospectus could offer: *'Practical Physiology* — This course is carried out in accordance with the regulations of the Royal College of Surgeons. There is a separate and well lighted room for microscopical work. A limited number of microscopes is provided, but it is very desirable that every student should possess an instrument of his own.' In following years each student was required to provide for himself 'a compound microscope with one or two eyepieces, an inch and 1/4 or 1/5 object glass'. Also he must provide for himself razor, pipettes, needles set in handles and slides and coverslips, because the students were now to prepare and mount tissues for themselves. Klein's *Manual of Histology* was recommended as a textbook for the practical classes.

Despite these reforms the teaching seemed inadequate in that too many students failed the examination at the College of Surgeons.

Yorkshire College of Science opened in 1874; as its interests widened to include Arts subjects the words 'of science' were omitted from the title. In 1876, by arrangement between the College and the Medical School, medical students were taught chemistry, botany and comparative anatomy in the College by the professors in the College; the Professor of Biology was Louis Miall (1842–1921), who was a very good zoologist, self-taught but elected FRS in 1892. As mentioned above, this freed rooms in the Medical School for practical classes in physiology (histology).

From its beginning Yorkshire College was concerned about the request by Owens College, Manchester, for university status and by the formation of Victoria University in 1880. To obtain similar status, both Yorkshire College and Leeds Medical School saw advantages in amalgamation and, after discussions through 1883, this was formally agreed in 1884. The enlarged Yorkshire College joined Victoria University in 1887 and the University of Leeds received its separate charter in 1904.

The amalgamation in 1884 provided for the first time the possibility of salaried lecturers in the Medical School and the first need was a Professor of Physiology. As part of the arrangements for amalgamation, an Endowment Fund for Physiology was set up, to which the Medical School contributed £1,000, and many of the senior medical staff made their private donations; the fund was sufficient for the Chair of Physiology to be advertised with a guaranteed salary of £300 per annum. It was perhaps fortunate that Wright had already served thirteen years as Lecturer in Physiology, was of high standing in obstetrics and could therefore be nominated as Professor of Midwifery and Obstetrics. Similarly Jacob was appointed part-time Professor of Pathology.

The new Chair in Physiology was advertised in 1884. At that time physiologists were being trained at four places in Britain — London, Cambridge, Edinburgh and Glasgow. There were several applicants of whom Birch (Edinburgh) was chosen in preference to North (p. 223).

de Burgh Birch (1852–1937)

Birch was the son of an officer in the Madras Medical Service and was educated in Switzerland and in Bristol where he was apprenticed to the practice of the hospital. In 1874 he began formal medical education in the University of Edinburgh, qualified MBCM in 1877 and proceeded to MD in 1880.

He apparently never practised medicine and did not go abroad but immediately after qualifying became assistant to Rutherford (p. 201) for some seven years, until he was appointed Professor of Physiology in Yorkshire College in 1884; this title became Professor in the University of Leeds in 1904 and he retired by age limit in 1917.

In Edinburgh he had shown promise as a research worker, published in 1878 a large paper in the *Journal of Physiology* on the histology of bone, but this seems to have been his only major original work. His promise as a research worker was unfulfilled; rather in Leeds he devoted himself to the organisation of a teaching department and to the general administration of the Medical School, Yorkshire College and the early days of the University of Leeds. He was Dean of the Medical School in 1900–07 and 1913–17.

At the time of Birch's appointment the Medical School was still in the building in Park Street and there was very little space for physiology, only a few microscopes and very little money to provide apparatus and other facilities. Birch evolved simple and effective student apparatus, which was made by Kershaw, a skilled mechanical worker.

Within three years of Birch's arrival it was obvious that the School was overcrowded and moves were begun to provide new accommodation, reaching the decision to build a new school on Mount Pleasant; a contract was agreed in 1891 and the new School was opened at the beginning of the session 1894–95. In this building Birch designed and equipped a Department of Physiology, which according to Raper (Birch's successor) was at the time the best in the provinces. In it Birch established full practical class teaching for medical students.

Birch joined the Physiological Society in 1892 but there is no record of communications to the Society by himself or assistants.

Alongside his work in physiology, Birch was active with the Volunteer Army, in which he reached the rank of Colonel. He formed a medical staff corps of which he was ADMS. This had been disbanded in 1914 but he was asked to reform it in 1915 and with it went to France. He was honoured by the award of CB.

In 1917 he retired to Bournemouth and died there in 1937.

During Birch's early years in Leeds there were several Demonstrators in Physiology, none of whom became professional physiologists. The first was

Alfred George Barrs (1853–1934). Barr's medical education was at Guy's and the University of Edinburgh; in 1875 he passed MB Edinburgh and took MD in 1882. He also passed MRCS, MRCP in 1884 and became FRCP in 1893. He began practice in Leeds as a physician particularly interested in cardiac diseases; was Assistant Physician in the Infirmary in 1884 and full Physician in 1892. In 1899 he was appointed Professor of Medicine and became a very highly respected figure in the College of Physicians. He followed the old custom in that, at the beginning of his career in Leeds, he became Demonstrator in Physiology but any wish for physiological work was lost in his later progress as a clinician.

Birmingham

John Berry Haycraft (1857–1922)

Haycraft was born in London and educated at Brighton Grammar School. His medical education was in Edinburgh where he graduated MBMCh in 1878. Ten years later he took the degrees of DSc and MD, with a Gold Medal for his MD thesis. After graduation he worked for a period in Leipzig with Ludwig, before he became Demonstrator in Physiology in Edinburgh under Rutherford. During this time he was elected FRS (Edinburgh) in 1880.

In 1881, Haycraft was appointed to a newly founded Chair of Physiology at Queen's and Mason Colleges, Birmingham, where his teaching attracted many students. However, in 1887 he was induced to return to Edinburgh to take charge of the department there during Rutherford's absence by illness. After Rutherford's return, Haycraft continued in charge of practical classes in Edinburgh, particularly an advanced course in physiology. Without any permanent appointment in Edinburgh, Haycraft in 1892 went to London as a research scholar of the British Medical Association.

He worked in University College and from this period a paper on 'A new Hypothesis of Vision' was communicated to the Royal Society of London by Professor Schäfer.

In 1893 a medical school began in Cardiff and Haycraft was

appointed as Professor of Physiology. He stayed in this post until 1920 and had much to do with the development of the School towards its reconstruction in 1922 as the Welsh National School of Medicine. Immediately, Haycraft had again the task of organising a Department of Physiology where none existed and by 1896 was publishing papers in the *Journal of Physiology* from the Cardiff Laboratory. After 1908, with the growing school, he began to insist on the need for a new building for physiology and his hopes were realised when a generous donor provided the means for the erection of a new and striking building, opened in 1913.

Haycraft suffered a hemiplegia in 1899 from which he made a remarkable recovery and he returned to his duties until forced by ill-health to make an early retirement in 1920. He was succeeded by Graham Brown. For the remaining two years of his life Haycraft worked in Langley's laboratory in Cambridge.

Throughout his career Haycraft was an active research worker publishing papers in the *Journal of Physiology* and other journals, the last posthumously. He joined the Physiological Society in 1891. His research covered many topics – the study of the striations of skeletal muscle, the chemistry of blood, amoeboid movement, coagulation of blood (especially the anticoagulant action of hirudin), the heart sounds. At that time his most widely known work was on the senses of taste and of smell, on which topics he contributed two chapters to Schäfer's *Textbook* of 1900. In 1894 he gave the Milroy Lectures to the College of Physicians on 'Darwinism and Race Progress'.

Haycraft was much appreciated as a founder of the Cardiff School, being regarded as a man of vision, determination and true science and withal a generous and honest man.

References

Anning, S.T. and Walls, W.K.J. (1982). *A History of the Leeds School of Medicine. One and a Half Centuries 1831–1981*. Leeds University Press.
Cohen, Lord of Birkenhead (1959). 'Richard Caton (1842–1926): pioneer electrophysiologist', *Proceedings of the Royal Society of Medicine*, **52**, 645–51 (P).
Kelly, T. (1981). *For Advancement of Learning: the University of Liverpool 1881–1981*. Liverpool University Press.
Thompson, J. (1886). *The Owens College: its foundation, growth and its connection with the Victoria University*, Manchester: J.E. Cornish.
Vincent, E.W. and Hinton, P. (1947). *The University of Birmingham; its history and significance*. Birmingham: Cornish Bros.

Biographies

Barrs: *Lancet* (1934), i, 546–7 (P).
Birch: *British Medical Journal* (1937), ii, 640.
            Proc. Roy. Soc. Edinburgh (1936–7), **57**, 402–3 (by Raper).
Caton: *British Medical Journal* (1926), i, 71–2.
            Cohen (1959) (P).
Gamgee: *British Medical Journal* (1909), i, 933–4.
            Lancet (1909), i, 1141–4 (P).
Haycraft: *Lancet* (1923), i, 158–9.
            British Medical Journal (1923), i, 86.
Jacob: *British Medical Journal* (1894), i, 611–12.
Priestley: *The Medical Directory, 1938*. London: Churchill.
Roberts: *British Medical Journal* (1899), i, 1063–6 (P).
Waters: *British Medical Journal* (1887), i, 246.
Wright: *British Medical Journal* (1908), i, 295 (P).

16

Zoologists and botanists

The developing subjects

In 1870, 'physiology' was still used as a rather general term, including within its ambit comparative anatomy, although with a bias towards functional interpretation (Ch. 8). In the immediately preceding chapters we have followed in Cambridge, University College and other centres of medical education, the development of one aspect of the general subject – the physiology of man and its investigation by experiments on vertebrates. This was the physiology relevant to medicine and the education of medical students, who were by far the largest group of students. After 1885 'physiology' came to be the title of a subject within universities, meaning the physiology of man as investigated particularly by experiments on mammals, and this meaning was unchallenged for the next fifty years.

Before 1885 the physiology of invertebrate species had become displaced from the stream of 'physiology', to form the new subjects of zoology and botany, in their modern definitions. After 1870 three great figures who carried physiology forward were Huxley, Foster and Burdon Sanderson and they were all general biologists; Burdon Sanderson was a very competent botanist and worked on carnivorous plants; Foster in Cambridge saw physiology as one branch of general biology, to be paralleled by functional aspects of zoology and botany. The conversion of zoology and botany into experimental subjects occurred in parallel with the use of experiments in medical physiology. Huxley's classes at South Kensington introduced the method of experiment into zoology and botany, as is shown in several of the biographies in this chapter.

The three new subjects of physiology, zoology and botany were about equally represented in the early membership of the Physiological Society. When it was founded in 1876, about one-third of the original members had careers in zoology or botany, about one-third in physiology and about one-third were practising physicians or surgeons. Lewes (p. 118), Huxley (p. 110) and Galton (p. 123) were senior biologists involved in the foundation of the Society and Charles Darwin and Carpenter (p. 107) were elected Honorary Members.

In the period 1870–85, most medical schools listed lectures in comparative anatomy and botany. Usually these lectures were not given by members of the clinical staff but by specialists. Where a medical school was associated with a university college, the teachers in zoology and botany often had the position of professor in the university. With physiology not yet a defined discipline, some of these teachers in biology carried out research of a physiological nature or were otherwise sufficiently known to the 'founders of physiology' to be invited to join the Physiological Society in the first years of its existence (p. 262).

The subjects of the following biographies were zoologists or botanists rather than what we would now call physiologists. They are included here because they were early members of the Physiological Society; one or two declined the invitation.

Zoologists

Sir Edwin Ray Lankester, FRS (1847–1929)

Ray Lankester was the son of Edwin Lankester (1814–74), who was Coroner for Central Middlesex, a worker in Public Health, much interested in all biological topics and particularly in the use of the microscope; he was one of the founders of the *Quarterly Journal of Microscopic Science*. Ray Lankester thus grew up in a home providing an intellectual, scientific background with eminent men, particularly Huxley, frequent visitors. His formal education was at St Paul's School, whence he obtained a scholarship to Downing College, Cambridge. In his first vacation he visited Oxford, was attracted by what he saw of Rolleston, and thereupon moved to Christ Church, Oxford; he obtained his Oxford degree in Natural Sciences in 1868 with first class honours, and was awarded a scholarship in Geology and a travelling fellowship.

In the next five years Lankester travelled widely in Europe, to Vienna to work with Rokitanski and to Leipzig with Ludwig. Moseley (p. 246) was his undergraduate friend and companion in many of these travels. Long periods were spent at the Stazione Zoologica in Naples, where he and Balfour (p. 247) were among the first to work with Anton Dohrn, its founder. While in Naples, Lankester had the pleasure of conducting Huxley to the top of Vesuvius and to Pompeii; Huxley was passing through Naples returning from a visit to Egypt.

In 1872 he had been elected Fellow and Tutor of Exeter College and was soon recognised as a good teacher. In 1874, aged only twenty-seven, he was appointed Jodrell Professor of Zoology at University College, London, succeeding Grant (p. 76). He stayed at University College for seventeen years until 1891, except for about a fortnight in 1882 when he was appointed Professor of Natural History in Edinburgh, found he did not like it and was allowed to resume his post at University College.

In 1891, in succession to his friend Moseley who had died, Lankester was elected to the Linacre Chair of Comparative Anatomy, Oxford and its accompanying Fellowship of Merton College. In addition in 1898 he became Director of Natural History at the British Museum. He resigned these posts in 1907.

Lankester's scientific work extended over the whole range of zoology. Beginning as a schoolboy, he worked and wrote quickly and prolifically throughout his life. Types mentioned in his obituary notices include protozoa, annelida, mollusca, arthropoda, medusae, rhabdopleura, amphioxus. Usually his work was precise anatomy but he also studied pigments in animals. He had learnt from his father skill with the microscope and in 1869 joined his father in the editorship of the *Quarterly Journal of Microscopic Science*, of which he became sole Chief Editor from 1878 to 1920. In addition to his vast output of scientific papers, books and lectures, he wrote some popular books on biological topics. He was elected FRS in 1875, was awarded a Royal Medal in 1884 and the Copley Medal in 1913. Lankester was much honoured by universities at home and abroad and was knighted on his retirement from the British Museum.

Above all, Lankester was a great teacher of the modern zoology; he organised museums at University College, London, at Oxford and the British Museum into effective teaching displays, and he established the experimental method in the teaching of zoology. In his undergraduate days, Lankester had learnt from Rolleston (p. 109) the value of skilfully organised museums. Also he had been associated with Huxley as Demonstrator in the biology classes at South Kensington (p. 113), and from these early contacts evolved his own powerful teaching. He was a superb lecturer, of great presence and powerful voice, who took great trouble to prepare illustrative material for each class. He was a very forceful personality, always encouraging to good workers but intolerant of shoddiness or pretence, about which he could express himself forcibly, leading sometimes to his being less than friends with other professors. He had enormous influence on his students both in London and Oxford. At University College, London he soon turned the run-down lecture course of Grant (p. 76) into lectures and practical classes of vivid attraction to students.

Such a man, continuing active until the age of eighty, was inevitably a leader in any biological development at the end of the nineteenth and beginning of the twentieth century. An important example was the formation of the Marine Biological Association with its laboratory at Plymouth. The idea originated with Lankester who was the active force in obtaining support. The Association was formed by a public meeting in London on 31 March 1884, at which Huxley, President of the Royal Society, was Chairman and referred all questions to Lankester; speakers in support were Carpenter (p. 107), Moseley (below), Bowman (p. 37) and Lankester. The officers of the Association included Huxley, Carpenter, Thiselton-Dyer, Foster, Moseley, Romanes and Burdon Sanderson with Lankester as Secretary, all members of the Physiological Society, and other members joined the Association at an annual fee of £1 1s or life membership of £15 15s (*Journal of the Marine Biological Association*, Vol. 1–2, 1887). The Physiological Society has since maintained its contact by an annual contribution and since the laboratory at Plymouth was opened in 1888, many physiologists have worked there. Lankester himself maintained an active interest in the laboratory but suffered a disability awkward to a marine biologist – he was sick every time the trawler left the calm waters of Plymouth Sound. For the same reason he refused invitations to visit America.

When the Physiological Society was formed in 1876, Lankester was Professor at University College, London and he was a founder member. He was a regular attender at meetings for about ten years; in December, 1876 the Society, with some doubts about the amount of the proposed work, supported his application for a grant from the government for research on the embryology of invertebrates and the natural history of the organisms concerned in putrefaction.

Henry Nottidge Moseley, FRS (1844–91)

Henry Nottidge, son of Henry Moseley, FRS (1801–72), a mathematician, was educated at Harrow before going to Exeter College, Oxford to read mathematics or classics. Instead he chose to study natural sciences under Rolleston and in company with Lankester, and attained first class honours in 1868. He and Lankester, close friends with similar travel grants, went together to Vienna, Leipzig and Naples, before Moseley in 1871 was appointed naturalist on the voyage of the *Eclipse* to Ceylon. On return from this expedition Moseley was next appointed botanist on the voyage around the world of *Challenger*. The voyage lasted four years and from it Moseley described and collected many species of animals and plants. He wrote an account of the voyage, *Notes of a Naturalist on the* Challenger.

Following the *Challenger* voyage he was elected Fellow of Exeter College and his work was recognised by election to FRS in 1879. In the same year he became Assistant Registrar of the University of London. Finally in 1881 he was appointed Linacre Professor of Comparative Anatomy, Oxford and Fellow of Merton College in succession to Rolleston, his friend and teacher. He became ill in 1887 and died in 1891.

In his short life and by his voyages he became a prominent zoologist, particularly as an authority on corals and arthropods. He was active in the affairs of zoology, including being a member of the Council of the Marine Biological Association, which he helped to found; he was always a close associate of his friend, Lankester. Moseley joined the Physiological Society in 1876, at its second meeting, but held no office; he resigned in 1889 when he had become ill.

Francis Maitland Balfour, FRS (1851–82)

One of the large family of James Maitland Balfour of Whittingehame, Haddingtonshire and Lady Blanche, daughter of the second Marquis of Salisbury, Francis Maitland was born in Edinburgh. Two of his brothers achieved fame as Conservative politicians. Arthur James (1848–1930) in the early stages of the 1914–18 war was a member of Lloyd George's inner cabinet and as Foreign Secretary (1916–19) was prominent in the Peace Conferences of 1919, issuing the 'Balfour Declaration'. He was also a philosopher and in recognition of his political career was created Earl of Balfour in 1922. A younger brother, Gerald William (1853–1945), also a Conservative politician, succeeded to the title, his brother having died unmarried.

As a boy Francis Maitland was introduced to natural history by his mother and soon became a keen geologist and was much interested in birds. At school in Hoddesdon and at Harrow, he made little of ordinary classes but benefited by the study of natural history as an extra subject with an interested master. In 1868 he was awarded a prize for an essay on the geology of East Lothian, so good that it was sent to Huxley for his judgement.

In 1870 he entered Trinity College, Cambridge at the time when Foster had just been appointed Praelector there. Balfour attended Foster's classes and decided to study animal morphology. He was noted by Foster as a good student with the usual result (p. 167) and Foster suggested that he should study the developing chick; thereby a great embryologist was started on his career. In 1873, Balfour was placed in second place in the Natural Sciences Tripos, being just edged out of first place by Newell Martin (p. 171). For the next year Balfour worked in Naples at the Stazione Maritima under Dohrn, with Lankester as his friend and companion.

He returned to Cambridge in 1875 to take over from Foster the lectures and practical classes in animal morphology and was appointed a lecturer of Trinity, although his lectures were given in the University Laboratory of Physiology and were open to the whole University. He was also actively researching, producing papers on animal embryology and he became a founder member of the Physiological Society. At this time he began his great treatise of *Comparative Embryology*, which when published in 1880–82 was immediately recognised as very great, embodying much of Balfour's own original work. He was elected FRS in 1878 and awarded a Royal medal.

His work was interrupted in 1882 by a serious attack of typhoid fever, contracted by helping a sick colleague on the Isle of Capri, and he only slowly recovered strength. In 1881 he was offered the Linacre Chair in Oxford, to which Moseley was appointed and also the Chair in Edinburgh, spurned by Lankester. He had resolved to stay in Cambridge, where in 1882 the University took the unusual step of electing him to a personal Chair of Animal Morphology. This he never occupied, because in the summer of 1882 he died tragically.

He had become a keen and proficient mountaineer, finding that such holidays helped his rather frail health. In the summer of 1882, he and a guide set out from Courmayeur to climb the Aiguille Blanche de Pétéret, failed to return, and it was found that, roped together, they had fallen from the mountain.

Balfour's work was continued by his pupil and friend, Sedgwick.

Adam Sedgwick, FRS (1854–1913)
Sedgwick's father was Vicar of Dent in Yorkshire and he was a great nephew of an earlier Adam Sedgwick (1785–1873), a great geologist and founder of the Sedgwick museum in Cambridge.

The second Adam Sedgwick entered Trinity College, Cambridge

in 1874 and passed First Class in Natural Sciences Tripos of 1877. He had come under the influence of Foster and Balfour and abandoned any thought of medicine, which he had first tried at King's College, London. From 1878 to 1882 he was Scholar and Fellow of Trinity and Demonstrator to Balfour, closely associated with his work.

On the death of Balfour, Sedgwick was appointed University Reader in Animal Morphology and Lecturer of Trinity, with the purpose of continuing Balfour's work. With the support of Foster and Lankester, Sedgwick developed the teaching in Animal Morphology to an important school, despite the presence of Newton (p. 165) with whom he maintained good relations. At the death of Newton in 1907 Sedgwick became Professor of Zoology.

Until 1897 Sedgwick was an active research worker on annelida and arthropodia; he was elected FRS in 1882. With some hesitation, in 1897, Sedgwick accepted appointment as Tutor of Trinity and thereafter was too fully occupied in administration and teaching for research to be continued. Finally in 1909 Sedgwick left Cambridge to be Professor of Zoology at Imperial College, London; he was wanted to reestablish the department originally started by Huxley.

Sedgwick was known in Cambridge as a loyal friend, but a blunt critic with a hasty temper, when he expressed himself in language forming Cambridge legends.

Although closely associated with Foster, Balfour and Lankester, Sedgwick did not become a member of the Physiological Society; possibly by his time zoology had become separated from physiology. Sedgwick's demonstrator was a close friend,

A. E. Shipley (1861–1927) who in 1894 was promoted to Lecturer in the Advanced Morphology of Invertebrates and in 1908 Reader in Zoology. Shipley was elected FRS in 1904 and was Master of Christ's College, 1910 to 1927. He, like Sedgwick, did not join the Physiological Society.

George John Romanes, FRS (1848–94)

Romanes was born in Kingston, Canada, where his father was Professor of Greek. His parents inherited a fortune and the family moved to London but were forced to reside abroad much of the time, owing to the boy's poor health. Consequently George Romanes had no regular school. In 1869 he entered Caius College, Cambridge to read natural sciences in which he obtained first class honours in 1870. Although he had no need to earn his living, he began a medical course but ill-health caused him to abandon this idea in 1872.

In 1874–76 he worked with Burdon Sanderson at University College, London, particularly on the excitability of nerves. He had a house at Dunskaith, Cromarty Firth and he there built for himself a marine laboratory for the purpose of studying starfish, medusae and sea urchins. Over the next ten years he observed the movements of these creatures in their natural environment and further studied them by experiments along the lines of those he had done with Burdon Sanderson. This was his most important research for which in 1879 he was elected FRS. For some of this work, Romanes cooperated with Ewart (see below) to produce their 1881 Croonian lecture on the locomotor system of echinoderms.

Romanes held no official appointments but always worked independently in matters of his own choice. He was Fullerian Professor at the Royal Institution in 1888. He took part in the formation of the Marine Biological Association in 1884 and was a member of its first Council (p. 246). In 1892 he became ill and died in 1894. A popular friend of all biologists, his early death was much lamented.

In 1876, when the Physiological Society was founded, Romanes was at University College and was present at the preliminary meeting at Burdon Sanderson's house. Yeo and Romanes were nominated joint secretaries and were members of the group which drew up the constitution. Later, in 1881, it was arranged that Romanes became treasurer and Yeo carried on the secretarial duties. Romanes resigned as treasurer in 1884, to be succeeded by Gaskell.

Romanes wrote on three subjects. Scientifically most important, were papers describing and studying the movements of sea animals; he gave two communications to the Physiological Society on this subject in 1880 and 1882. A second subject was 'Animal Intelligence'. In 1878 he gave a lecture to the British Association on 'Mental Evolution in Animals'. Thirdly he was interested in the mechanism of evolution, writing a book, *Darwin and after Darwin*, in 1892. He had met Darwin as a young man and had become a personal friend.

Romanes began with deep and orthodox religious convictions, considered taking Holy Orders and wrote a prize essay on 'Christian Prayer considered in relation to the belief that the Almighty governs the world by general laws'. As his scientific life advanced, he veered towards agnosticism but towards the end of his life, he again accepted religious orthodoxy.

In 1890 he went to live in Oxford, where Burdon Sanderson was then Professor of Physiology. He used part of the Romanes fortune to endow an annual Romanes lecture in the University of Oxford; the first three lecturers, nominated by Romanes himself, were Gladstone, Huxley and Weismann.

James Cossar Ewart, FRS (1851–1933)

Ewart was born at Penicuik, Midlothian and was educated there until he entered the University of Edinburgh in 1870 as a medical student; he qualified BMMS in 1874; and would therefore have studied physiology under M'Kendrick acting as deputy for Hughes Bennett (p. 98).

For a short period Ewart was Demonstrator in Anatomy at Edinburgh, before he went to London and was appointed Curator of the Zoological Museum at University College. Lankester (p. 244) had just been appointed Professor and was developing the museum and Ewart also assisted Lankester in the newly introduced practical classes in zoology.

In 1878 Ewart returned to Edinburgh as Lecturer in Anatomy at the Extramural school but was also immediately appointed Professor of Natural History at Aberdeen. He stayed in Aberdeen only four years; in 1882 he was appointed Professor of Natural History in Edinburgh where he remained until he retired in 1927.

A link with physiologists came with his first marriage; he married the sister of Edward Schäfer (p. 147). She died leaving one daughter and Ewart married again, twice. Throughout his life Ewart was devoted to his birthplace, returning to live in Penicuik while he was professor in Edinburgh and died there in 1933.

Ewart was an active research worker. In London he worked on the structure of the eye, on the lamprey, on the fertility of deer and also on bacterial organisms; his thesis for MD (Edinburgh) was on *bacillus anthracis*. In Aberdeen he became involved in marine biology and had a small research station where he studied matters connected with the fishing industry and also worked with Romanes to produce their joint Croonian Lecture of 1881 on the locomotor system of echinoderms. This work was continued in Edinburgh where he also studied the electric organ of the skate and the sense organs of elasmobranchs. Later he moved away from sea animals to study the development of the skeleton in horses and the growth of feathers in birds, particularly using penguins.

Ewart became best known by his work on animal breeding begun about 1895. His first experiments were on the idea of telegony, that a female mated with an individual male would remain 'infected' with the characteristics of that mate. Ewart mated a mare with a zebra, producing striped offspring known as 'Tartan cuddies', but the progeny of subsequent mating of the mare with horses had no stripes. The results were published as the *Penicuik Experiments* in a book with striped covers. After 1913, on a University farm at Fairslacks, Ewart led experiments on the breeding of sheep, supported by a Committee on Animal Breeding of the Board of Agriculture of Scotland. In connection with this work he visited Australia and New Zealand in 1923.

Ewart was elected FRS in 1893 and in 1907— 09 he was a member of the Council and in 1908—09 Vice-President. Elected a member of the Physiological Society in 1877 at the first annual general meeting, he attended some meetings. At a meeting in Edinburgh in 1905 he is noted as a guest; presumably he had resigned as his interests went away from physiology.

Revd Charles John Francis Yule (1848—1905)

Yule was the son of a naval officer, born at East Stonehouse, Devon. He went to Magdalen School, Oxford and in 1868 matriculated into Balliol College, Oxford. However in 1869 he migrated to St John's College, Cambridge and obtained first class in Natural Sciences Tripos, 1872, and was elected a scholar of St John's. He took his BA (Camb.) in 1873.

Then, after a brief period as assistant master at Magdalen School, he became a Fellow of Magdalen College, Oxford and took MA Oxford in 1875. For the next ten years Yule filled a succession of posts at Magdalen. From 1873—84 he was Tutor in Natural Sciences and pioneered the teaching of biology and physiology to Magdalen students. In 1876 he was junior Dean of Arts, in 1878 Junior Bursar, in 1880 Vice-President and in 1882 Senior Dean.

In 1885 Yule was ordained in Worcester Cathedral and served in several parishes. He was curate of Alcester 1885—89, vicar of Horspath, 1889—92 and of Ashbury, 1892—1900 before he retired to Eynsham in 1900.

He became a member of the Physiological Society at its foundation in 1876 and attended a few meetings in the next few years, but made no communications.

Botanists

Sir William Turner Thiselton-Dyer, FRS (1843–1928)

Thiselton-Dyer was born in London and educated at King's College School, before entering King's College, London in 1861 as a medical student. He did not complete the course; instead he went to Christchurch, Oxford in 1863 to read mathematics and obtain second-class honours in 1865. He then read natural sciences under Rolleston (p. 109) with Moseley and Lankester (also undergraduates at that time); he obtained first class honours in 1867.

Immediately he became Professor of Natural History at the Royal Agricultural College, Cirencester. He took the degree of BSc (London) and in 1870–72 was Professor of Botany at the Royal College of Science in Dublin. His next appointment was Professor of Botany at the Royal Horticultural Society at South Kensington, where he made two important contacts. Firstly he met Huxley and became Demonstrator in Botany to Huxley's course and practical classes in general biology. After 1874 Thiselton-Dyer gave the lectures in botany in these classes, with Vines as Demonstrator.

The second contact determined the rest of his life. He met Sir Joseph Hooker, Director of Kew Gardens, whose daughter he married. For about a year, Thiselton-Dyer acted as secretary to Hooker, until in 1875 he received an official appointment as assistant director. This post had lapsed but was re-established for him. He was made responsible for dealing with the colonial problems which were referred to Kew and so played an important part in such activities as introducing into Ceylon cacao plants from Trinidad; and into Ceylon and Malaya rubber plants from South America, with great commercial success. Hooker resigned in 1885 and Thiselton-Dyer became Director of Kew. He continued Hooker's work in developing the garden; he was responsible for a very precise organisation of the working of the garden and introduced the uniformed attendants patrolling in the open hours. More importantly Thiselton-Dyer developed the Jodrell Laboratory at Kew, providing for investigation of the physiology of plants and the advanced education of botanists. He was something of an autocrat in the management of Kew but was much respected and was an important factor in the new, practical botany of his time. He was elected FRS in 1880 and was knighted in 1899.

In 1905 he retired from all scientific work to live the life of a country gentleman in Gloucestershire. He was an original member of the Physiological Society and for some years occasionally attended its meetings; he held no official posts in the Society.

Sir Francis Darwin, FRS (1848–1925)

Francis Darwin was the third son of his famous father, and is noted as his biographer. *The Life and Letters of Charles Darwin*, published in 1887 is an excellent biography; a shortened version, *Charles Darwin*, appeared in 1892.

Francis was born at Downe and lived there until his father's death. He entered Trinity College, Cambridge to pass first class in Natural Sciences in 1870. He then studied medicine at St George's Hospital and qualified MB in 1875 but never practised. A first published paper, while still a medical student, described the post mortem findings in an ostrich which had died of copper poisoning after swallowing 'two pennies and fifteen halfpence'. His MD thesis in 1876 was on inflammation and reported work done with Klein at the Brown Institute (p. 136); Klein appears to have influenced him away from clinical work towards science.

Then for eight years he lived at Downe as secretary to his father and assisted him with experiments on plants; they collaborated in a book on movements in plants. In 1882, Francis was elected FRS.

After the death of his father in 1882, Francis Darwin settled in Cambridge with a Fellowship of Christ's College; he was appointed University Lecturer in Botany in 1884, and became Reader in Botany, 1888–1904. He did much to develop the teaching in botany, especially in 1892–95 when he was deputy for Babington (p. 165). He wrote a book for medical students *The Elements of Botany* and with E.H. Acton a manual, *The Practical Physiology of Plants*. He worked experimentally on plants, particularly on transpiration and the uptake of water, using ingenious apparatus designed and made in collaboration with his brother Horace (p. 166).

In the later period of his life he was a member of the Council of the Royal Society, its foreign secretary in 1903–07 and Vice-President in 1907–08. He was knighted in 1913. He was also President of the British Association. At the foundation of the Physiological Society in 1876, Charles Darwin was made an Honorary Member and Francis was a foundation member, who occasionally attended meetings until his resignation in 1888. His resignation is perhaps an indication of the drift of botany to a subject separate from physiology.

Sidney Howard Vines, FRS (1849–1934)

Born in London, Vines entered Guy's Hospital to study medicine but gained a scholarship by which he went up to Christ's College, Cambridge in 1872. He graduated in Natural Sciences Tripos in 1875 with first-class honours in Botany; one of his teachers had been Michael Foster and he was also encouraged by Newell Martin, a little senior to him in Christ's College.

As a medical student in London and an undergraduate at Cambridge, Vines was a demonstrator in Huxley's classes of general biology at South Kensington. When the botany lectures in these classes were taken over by Thiselton-Dyer, Vines continued as his demonstrator.

By his contacts with Foster, Martin and Huxley, Vines's interests were directed towards the physiology of plants and he went to Würzburg to study with Sachs. On returning to England Vines taught plant physiology in Foster's laboratory as part of the course in physiology (p. 164). He also carried out research in plant physiology and about 1878 published papers in 'Studies from the Physiological Laboratory in the University of Cambridge' (p. 165) and in the first volume of the *Journal of Physiology*.

The teaching of botany was being reorganised in Cambridge and in 1883 Vines was appointed Reader in Botany and a laboratory for Botany was established; at this time Francis Darwin was also Lecturer in Botany. The Professor, Babington (p. 165) was then a sick man, the work was done by his deputies, and modern teaching with practical classes resulted.

In 1888 Vines was appointed Sherardian Professor of Botany at Oxford and held this chair until his retirement in 1919. He did much to institute modern teaching of botany in both Cambridge and Oxford. He wrote a *Course of Practical Instruction in Botany* (1888) and *Lectures on the Physiology of Plants*, published in 1886. He was elected FRS in 1885 and became President of the Linaean Society in 1900–04. From his retirement at Exmouth, he continued to take an interest in botanical matters after 1919.

He was elected a member of the Physiological Society at the first annual meeting in 1877 and was for the next ten years occasionally present at meetings.

George Thomas Bettany (1850–91)
Born at Penzance, the son of a schoolmaster, G. T. Bettany was
apprenticed to a local surgeon in 1865–68 and from 1868–70 he was
a student at Guy's Hospital in London, gaining the degree of BSc
London in 1871. Apparently he never proceeded to obtain medical
qualifications.

In 1870 Bettany was admitted to Gonville and Caius College,
Cambridge, where he was a scholar 1871–77. He took first class in
Natural Sciences Tripos in 1873 and became BA in 1874, MA in 1877.
He was Lecturer in Botany in two places – Guy's Hospital 1876–86
and at Girton and Newnham in Cambridge 1876–78. At this time,
1877, he joined the Physiological Society and was a frequent attender
at meetings until he resigned in 1887. He wrote no papers and
apparently did no research, but wrote students' primers in botany and
physiology.

Progressively he changed to being a writer on subjects associated
with science. He wrote a biographical history of Guy's Hospital (1892)
and a life of Charles Darwin. Of particular interest to physiologists
is *Eminent Doctors: their lives and their work*, published in 1885 and
cited in Part I of this book. Other books included *The World's
Inhabitants: animals and plants; Teeming Millions in the East*, and
World Religions.

References

Biographies

Balfour:	*Proc. Roy. Soc. London* (1882), **35**, xx–xxvii (by Foster).
Bettany:	*Alumni Cantabrigienses* (J. A. Venn, 1940), Part II, Vol. I. Cambridge University Press.
	Modern English Biography, Vol. IV (F. Boase, 1965). Cass & Co.: London.
Darwin, F.:	*Nature* (1925), **116**, 583–4.
	Proc. Roy. Soc. London B (1932), **110**, i–xxi (P).
Ewart:	*Nature* (1934), **133**, 165–6.
	Obit. Fellows of Roy. Soc. London (1934), **1**, 189–95 (P). (by F. H. A. Marshall)
Lankester, E.:	*Lancet* (1874), ii, 676–7.
Lankester, Ray:	*Proc. Roy. Soc. London* (1930), **106**, x–xiv (P).
	Nature (1929), **124**, 309–14, 345–7.
Moseley:	*Nature* (1891), **45**, 79–80 (Lankester).
Romanes:	*Proc. Roy. Soc. London* (1894–5), **57**, vii–xiv (Burdon Sanderson).
	Nature (1894), **50**, 108–9 (by Lankester).

Sedgwick: *Proc. Roy. Soc. London B* (1912–13), **86**, xxiv–xxix.
Shipley: *Proc. Roy. Soc. London B* (1928), **103**, i–viii (P).
Thiselton-Dyer: *Proc. Roy. Soc. London B* (1930), **106**, xxiii–xxix.
                  Nature (1929), **123**, 212–5.
Vines: *Obit. Fellows of Roy. Soc. London* (1934), **1**, 185–8 (P).
                  Nature (1934), **133**, 675–7.
Yule: *Alumni Cantabrigiensis* (J. A. Venn, 1954), Part II, Vol. VI.
                  Cambridge University Press.

CONCLUSION

17

Physiology in Britain in 1885

Organisation and teaching

By 1885 physiology had become established as a separate subject. By lectures and practical classes it was being taught to medical students and to a few science students, much as it was to be taught for the next sixty years, until the revolution in methods produced by a new technology of electronics.

The General Medical Council now recognised physiology as a subject essential to proper medical education and courses in practical physiology were required by universities and by the Colleges of Surgeons, Physicians and Apothecaries which granted registrable medical qualifications. There were separate examinations in physiology with questions requiring knowledge of experimental physiology.

University and King's College in London and the Universities of Cambridge, Oxford and Edinburgh, Glasgow and Aberdeen now had schools of physiology. In these institutions there were professorships in physiology, with a staff of lecturers, assistants and technicians manning practical laboratories for students at all levels. In the Medical Schools of the London Hospitals, the change from clinicians who taught physiology to full-time professional teachers had been achieved only at St Mary's but the pattern was established, although it was not completely achieved for many years. In the provincial medical schools in 1885, there were endowed chairs of physiology at Manchester, Liverpool, Leeds and Birmingham, and soon other schools followed this pattern, which was a necessary step as a provincial school joined a local embryo university to seek a charter to grant degrees, which formed a registrable medical qualification.

Physiologists now had something to teach. In successive editions of Foster's *Textbook* (p. 31), Kirkes's *Handbook* (p. 30) and in Yeo's *Manual* of 1885 (p. 189) it can be seen that physiological knowledge had reached a state in which it formed the foundation of physiology as taught for the next fifty years. Physiology had now cast 'general anatomy'; it relied on accurate histological description but, in his textbook, Foster could assume that his readers had knowledge of histology. Over the following fifty years the teaching of histology moved from physiology back to anatomy, but the microscope remained an important tool of the physiologist. Pathology had become a recognised separate subject and was taught independently of physiology. There was now effective knowledge of the chemistry of body substances and most schools had a member of staff who specialised in the teaching of physiological chemistry, although biochemistry did not become recognised as a separate subject until about 1900. The main metabolic processes in the body were known.

Research work

Measurements of body function had now become the basis of physiology. Methods of measuring substances in body fluids had developed with knowledge of their chemistry. Methods were being developed to measure oxygen and carbon dioxide in expired air, making the beginnings of the measurement of metabolic rate. The kymograph was a usual piece of equipment and the mercury manometer provided measurement of blood pressure; oncometers were used to measure the volumes of organs and levers to record muscular contraction. Mostly these methods had been introduced in continental laboratories but British physiologists such as Roy (p. 183) were developing them into methods used by circulatory physiologists for the next fifty years.

Most importantly, with increasing experience of anaesthetics, the anaesthetised cat, dog or rabbit had become a standard physiological preparation. Until 1876 the use of anaesthesia by physiologists was by no means universal and their experiments were thus extremely limited. However, after the Royal Commission and the Vivisection Act of 1877, the use of anaesthesia was developed by the young generation of experimenters, such as Schäfer, Gaskell and Langley, until by 1885 its use was universal. Objections to the use of cats and dogs were thereby removed and the experimental methods to be used by such great experimenters as Sherrington, Bayliss, Starling were established before 1885 by those we have here called founders of physiology.

The physiology laboratories at University College and King's College in London and at Cambridge, Oxford and Edinburgh were active research institutions and there were also other laboratories, such as the Brown Institute, producing papers for publication and training the next generation of experimental physiologists. The colleges of Cambridge provided fellowships and scholarships for research workers and similar scholarships came from sources such as the BMA, the Royal Society and endowed funds. After 1885 young workers could usually find sufficient financial support and it was no longer necessary for a man interested in physiological research to begin by depending on private income or the profits of clinical work for his apparatus and livelihood.

Journal of Physiology

In 1885 there were sufficient original papers on physiology to maintain its own journal. The *Journal of Physiology* had been founded in 1878 by Foster, with the help of Dew-Smith (p. 181), and immediately became the main place of publication of original work in physiology; its early numbers show the sort of work done by physiologists about 1880. It included some papers on botany and some on invertebrate animals but the majority of papers were on man and vertebrates. Foster's choice of co-editors – Burdon Sanderson, Gamgee, Rutherford and the Americans, Bowditch, Martin and Wood – shows that he regarded physiology as mainly experimental work on vertebrates and man, for this was the sort of work in which these co-editors were already expert.

The Physiological Society

Another sign of the maturity of Physiology was the formation in 1876 of the Physiological Society. This also indicated the increasing importance of experimental work on vertebrate animals, because an immediate object of the founders was to provide a forum and a concerted force to influence the impending legislation on vivisection. The formation of the Society was recorded by Sharpey-Schafer (1927), who was himself a founder member, then junior to the main founders.

Society members later accepted that two main founders were Foster, then Praelector in Physiology at Trinity College, Cambridge and Burdon Sanderson, then Jodrell Professor at University College, London. On Sanderson's death in 1905, a minute of the Society recorded 'the loss of its oldest and most distinguished member' and in 1907 the annual report recorded the death of its 'chief originator, Sir Michael Foster'. The actual first step was a meeting at Burdon Sanderson's house for which invitations were signed by Burdon Sanderson and written out by his wife. The nineteen who attended were: Burdon Sanderson, Sharpey, Huxley, Foster, Lewes, Galton, Marshall, Humphry, Pavy, Lauder Brunton, Ferrier, Pye-Smith, Gaskell, M'Kendrick, Klein, Schäfer, Francis Darwin, Romanes and Yeo. Lewes acted as secretary for this meeting, at which a committee was chosen to prepare a draft constitution and Romanes and Yeo were asked to act as secretaries.

The next step was the drawing up of a list of thirty-seven names of men who were thought to have an interest in physiology and were sent an invitation to become founder members; thirty-five accepted. Of the thirty-five, ten were actively teaching physiology, mainly to medical students in departments which had a Professor of Physiology – Cambridge; University College, King's College, London; Manchester; Edinburgh – and where experimental work on animals was then in progress. They may be regarded as the central core of professional physiologists to whom the formation of the Society was an urgent necessity. They were Foster, Gaskell, Langley, Martin (Cambridge), Burdon Sanderson, Schäfer (University College), Yeo (King's College), Gamgee (Manchester), M'Kendrick (Glasgow), Rutherford (Edinburgh). They were all medically qualified.

Of the original thirty-five, another twelve were surgeons or physicians who were then, or had been, part-time lecturers in physiology in a medical school; they were medically qualified and in their later years abandoned physiology to teach medicine or surgery. They were – Brunton, Caton, Ferrier, Humphry, Klein, Marshall, McCarthy, McDonnell, Pavy, Pye-Smith, Power and Richardson. Thus twenty-two of the thirty-five men regarded in 1876 as 'physiologists' were connected with medical teaching and would have a bias towards experiments on man or vertebrates.

The remaining thirteen were general biologists. Older men who were active in the formation of the Society were Huxley, Lewes and Galton, and Romanes became, with Yeo, a secretary of the new Society. Presumably they felt sufficient common interest with the older medical founders (Foster, Burdon Sanderson, Pavy) to join with them. This common interest seems not to have been universal; other older men, Carpenter and Rolleston, refused to join, possibly because of lack of common interest but possibly also because of uncertainty about the ethics of animal experiments. Many younger comparative anatomists and botanists did join – Balfour, Lankester, F. Darwin, Dew-Smith, Thiselton-Dyer, Ewart, Yule – in sympathy with their teachers and more senior colleagues in University College and Cambridge. Equally we find in Chapter 16 that others, Shipley, Sedgwick did not join the Society.

The scope of physiology

The original membership of the Physiological Society reflects two influences which operated in the development of the subject after 1870. On the one hand there was the continuing influence of medicine; throughout this book physiology developed because it had to be taught to medical students. A new influence after 1870 was the teaching by practical classes, which can be ascribed in England to the influence of Huxley's practical classes in biology (p. 113); many of the physiologists in England had been demonstrators in these classes. The other group were biologists, particularly zoologists, who had no connection with medicine.

The founders of physiology, and of the Physiological Society, were of these two groups. Throughout this book most of the biographies are of medical men and, as stated above, twenty-two of the thirty-five original members of the Society were medically qualified and connected with the teaching of medical students. Sharpey was elected an Honorary Member as the revered figurehead of this group. The second group of thirteen original members were general biologists, later to become zoologists or botanists, and Charles Darwin was elected an Honorary Member as the senior living biologist.

In University College, London, the two strands developed side by side after 1874, with Lankester, Professor of Zoology and Burdon Sanderson, Professor of Physiology. At Oxford, with no effective medical school before 1885, Zoology grew under the influence of Rolleston and then Moseley. Foster went to Cambridge and was the physiologist in the medical school being rejuvenated by Humphry and George Paget. He also had the idea of physiology as the central subject for all biological sciences and instituted classes in a modern zoology and botany within the ambit of his physiological laboratory; these classes overreached those of the elderly Professors of Zoology and Botany. By 1885 teaching in comparative anatomy was by Balfour and Sedgwick and in botany by Vines and F. Darwin, all pupils of Foster, while the teaching of physiology was now largely by Gaskell and Langley.

Progressively areas of study became more distinct in their outlooks and with the organisation of universities became separate departments of zoology, botany and comparative anatomy. In the medical field the study of diseases formed a separate subject of pathology and men such as James Paget, Simon, Beale and Roy are to be regarded as founders of pathology. By 1885 physiology was left as the study of function in man and vertebrate animals, with strong medical links and with experiments on anaesthetised animals as its principal experimental method. This crystallisation was perhaps an indirect result of the antivivisection campaign and legislation.

Physiology thus defined and organised was to make enormous strides in Britain over the next thirty years. The period 1885–1915 can well be called the Golden Age of British Physiology, when it led the world. Its active men were nearly all medically qualified. To the younger founders, such as Schäfer, Gaskell and Langley, who continued as active research workers, were soon added the great names of Sherrington, Haldane, Bayliss, Starling and many others. Their great achievements were based on foundations laid by *Founders of British physiology, 1820–1885*, but particularly the immediate founders of 1870–85.

Reference

Sharpey-Schafer, E. (1927). *History of the Physiological Society during its First Fifty Years 1876–1926*. Issued as a Supplement to *Journal of Physiology*, December.

Index

Bold type indicates main biographical notes.
* indicates member of the Physiological Society before 1880.